Reflections on the Technological Society

Egbert Schuurman

D1741795

Wedge Publishing Foundation/Toronto

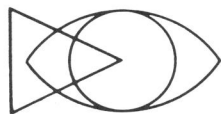

ISBN 0-88906-150-5

Originally published as follows:
"Between technocracy and revolution": "De kulturele spanning tussen technokratie en revolutie," translated by Harry Van Dyke
"Reflections on the technological-scientific culture": "Na-denken over de technisch-wetenschappelijke kultuur," translated by Lammert Tenyenhuis

Funds for the translation of these essays were provided by the Free University in Amsterdam

Design: Anthony Goodhoofd Associates

Wedge Publishing Foundation
229 College Street, Toronto, Canada M5T 1R4

Printed in Canada

Contents

Preface

Western culture, characterized by science and by modern technology, is in a crisis. This is shown not only by the philosophical reflection on our culture, but also by the many problems facing our culture.

In this brief volume I have collected a number of lectures that attempt to show the spiritual and historical backgrounds of that crisis. These backgrounds are generally ignored, yet they must be examined if we are to find a meaningful perspective for the future.

The first essay is a slightly revised and expanded version of my inaugural lecture at the Technological University of Eindhoven in the fall of 1973; it analyzes the tension in our culture between technocracy and revolution. The second essay deals particularly with the spiritual roots of our environmental crisis. The third essay is a revised version of my inaugural lecture at the technological University of Delft, fall 1975; this essay consists of a critical analysis of the relationship between science and culture.

My aim throughout these essays has been to reduce the dilemma facing our civilization to its basic elements. This will prevent rushing into matters of detail and put us in a position to pose the right questions. It will also stimulate us to seek the only sure basis on which we can begin to solve a major issue that plagues our culture.

The many footnotes documenting my arguments have been reduced drastically. The works I relied on chiefly are listed in a Select Bibliography.

I thank Harry Van Dyke and Lammert Tenyenhuis for the care with which they have translated chapter one and chapters two and three respectively.

Egbert Schuurman
Breukelen, The Netherlands
Spring, 1976

Foreword

Dr. Egbert Schuurman is professor of christian philosophy at the Delft and Eindhoven Institutes of Technology. He was appointed to these two positions by the Foundation for Calvinist Philosophical Education, which avails itself of the legal privilege of private associations to appoint professors and lecturers of their choice in special chairs at state universities and comparable centers of higher education in The Netherlands. Dr. Schuurman also teaches philosophy of culture at the Free University of Amsterdam. Prior to pursuing his doctoral program in philosophy, he completed a graduate program in mechanical engineering. His major publication is *Technology and the Future: A Confrontation with Philosophical Views* (1972) which is being translated and forms the basis for the chapters in this booklet.

Schuurman discusses technology in the context of the christian philosophy that finds its roots especially in the thought of Herman Dooyeweerd (1894-1977). He thus continues for our generation what his mentor, Professor Hendrik van Riessen, began immediately after the second world war, namely, an analysis of the place of technique in western civilization in the light of christian assumptions. This analysis thus runs parallel with the work of two fellow Christians — Jacques Ellul, the French thinker who published *The Technological Society* in 1954, and George Grant, the Canadian philosopher whose *Technology and Empire* appeared in 1969. One way of introducing Schuurman to the English-speaking world is by comparing him briefly with Ellul and Grant.

With Ellul, Schuurman rejects the positivist's and pragmatist's adoration of technology as the neutral tool by which we can create the great and global society of the future, in which all of mankind's needs supposedly will be met by the inexhaustible resources of an industrial-technological apparatus. Further, Schuurman accepts Ellul's diagnosis of the cause of the destructive derailment of technology in contemporary culture. Both single out the notion of the autonomy of technology — as if it were a law unto itself, not subject to a norm outside itself — as the immediate cause of that derailment. But at this point Ellul and Schuurman part ways. Ellul seems to say that the very nature of modern technology implies its autonomy. For him technology is basically evil and inhuman. Moreover, he does not search for a deeper source of the notion of autonomy. Schuurman, in contrast, argues that autonomy is not inherent in technology but that it is the religio-spiritual assumption of post-medieval modern man.

Here Schuurman and George Grant present parallel diagnoses. Ever since his rejection of the religion of progress in the early sixties, Grant has forcefully argued that western technology is rooted in the main heresy of the modern age, namely, the belief that man's essence is his freedom, his autonomy. In view of this both Grant and Schuurman point to the weakness of the counter-cultural critique of technology because that critique is still founded on the ideal of absolute human freedom. After all, that ideal gave rise to western man's mastery of nature by science, technology, and industrial production. In modernity, freedom means mastery. Thus both Grant and Schuurman realize that within modernity's dialectical swing from freedom to mastery and from mastery to freedom it is impossible to overcome the autonomy of technology.

If there is then so much similarity between the diagnoses of technology that Grant and Schuurman present, why should one bother to translate and read Schuurman? Because, while Grant suggests that in our coming to grips with the error of modernity we should painstakingly recall the Greek view of nature and man's place within it, Schuurman suggests that we listen to the biblical revelation of reality as creation and man's place within it. While for Grant and other representatives of the neoclassical school of cultural reflection — like Eric Voegelin — classical philosophy and biblical revelation are correlative and complementary, for Schuurman they are to be carefully distinguished. For Grant the autonomy of modern science and technology presupposes the rejection of the Greek view of nature; for Schuurman it presupposes the rejection of the biblical view of creation.

This means that Schuurman not only can present a critique of an *autonomous* unfolding of science and technology but also can point to the foundation of their *responsible* disclosure, even at this late hour in the disintegration of western society. For the revelation of creation, known to us in Jesus Christ, presents the dynamic, normative guidelines for the execution of man's task in history, including the structurally limited but nonetheless positive service required of science and technology. In his critical appraisal of the role of autonomous science and technology, Schuurman therefore calls for the recovery of the consciousness of created reality as the proper foundation for science and technology. That call is imperative in our era of gnostic futurisms, in which the *alpha* of creation has been repudiated to make possible the realization of a secular *omega*, an immanentized eschaton.

Bernard Zylstra
Institute for Christian Studies
Toronto

Between technocracy and revolution

It has been estimated that perhaps eighty percent of all the scientists and engineers who ever lived are living today. A figure like that makes one realize the exceptional character of our time. A growing number of people have jobs related in one way or another to the development of science and technology. In addition they find themselves surrounded more and more by the products of technology, both on the job and at home. We live in a technological civilization that is growing all the time in strength and scope and that is spreading across the globe.

Until recently this development was generally applauded as the surest way to progress. Today, however, as man turns his mind to the future, he sees enormous problems ahead. These problems seem to have been produced by the close cooperation between science, technology, economics, and politics. Confronted by the growing perplexities and conflicts in society and by the threat of ever greater catastrophies, some people have lost their enthusiasm for technological advancement. The destruction that man has brought about and continues to bring about with the aid of technology is so terrifying and seems so irreparable that it looks as if even man himself may eventually become a victim of technology.

A tension has thus arisen between, on the one hand, the seemingly anonymous, impersonal, and objective development of technological-scientific power and potential and, on the other hand, the subjective, personal decisions of human beings who contribute to this development. It is a tension that plagues those who are involved in this process; yet this same tension has an even greater effect on people who do not directly contribute to science and technology, but who may well be swept along by the avalanche of their cumulative applications as they give way to a spirit of apathy or anxiety, unbearable tension, alarm, or even panic.

These growing misgivings notwithstanding, there are still many technologists, engineers, and technicians who champion the unremitting growth of technology. Today's problems, they say, are the problems of a technology in its infancy; they can be solved by exploiting more fully the possibilities of technology. The method of

1

technology should be extended to other areas, such as economics and politics. What is good for technology is good for all culture. As technology advances, so will our society as a whole. The people who argue this way are called technocrats. Looking at the possibilities of technology, they are as optimistic about the future as ever.

Meanwhile the view of the technocrats only intensifies the reaction of those who take a radically different stance: the revolutionary utopians. Their assessment of our situation today is very somber. Looking at present trends in technology and at the growing power of science over our lives, they are pessimistic about the future and prefer to stake their hopes on alternative plans for tomorrow's society. They dream of a utopia in which man lives a free, happy, and carefree life. This utopia is to be realized by overturning our technocratic society through revolution. Obviously the realization of these ideas would have far-reaching consequences for the further development of technology and for its possible role in economics and politics.

The field of tension in present culture is governed by the two poles of technocracy and revolutionary utopianism. These constitute the contrasting orientations or twin foci of the struggle that is being waged in modern culture. In this essay I cannot discuss all the intermediate positions that have been or could be proposed.[1] Therefore when I refer to specific philosophers and futurologists to illustrate the spirit of technocracy or of revolutionary utopianism, I do not mean to imply that their thought coincides nicely with either one of these cultural currents; my point each time is that the pattern of thought in question contains elements that confirm the contrary tendencies in our culture. Precisely because the issue is one of basic "orientations," one rarely runs into a thinker who is a strict technocrat or a strict revolutionary utopian. In most cases, however, a man's thinking clearly gravitates towards either the one pole or the other and unmistakably reflects its specific temper.

In this essay I shall investigate in what respects technocrats and revolutionary utopians differ, what kind of future each group pursues, which methods and which criteria each uses, and how each judges the other.

To gain a deeper insight into the cultural conflict I have outlined, I shall deal briefly with the historical background of the intellectual and spiritual mainsprings behind the rise and development of science and technology in western civilization. Such a historical analysis, highly desirable in itself yet woefully absent from most treatments of the topic, should demonstrate the contemporary relevance of a christian philosophical framework. The relevance of a christian vision must not, however, stop once the critical evaluation has been made; it should also indicate concretely the direction in which we might seek a solution. That is what I propose to do;

however, the scope of the present essay leaves me room for a perspective only. I hope this perspective will show in broad outline how we can break through the cultural tension between technocracy and revolution and rise above the dilemma with which the current debate seems to saddle us.

The technocrats

The technocrats — also known as the ideologists of planning — include such men as Herman Kahn, Antony Wiener, Olaf Helmer, Karl Steinbuch, Erich Jantsch, and the marxian philosopher Georg Klaus. To a man, this motley group whole-heartedly approves of modern science and technology. Technology is for them the motor of progress, and scientific discoveries are the fuel. These people have set their hopes in particular on the development of the computer in close connection with systems theory and cybernetics. The computer is regarded as a mighty tool with which to investigate, guide, and control the future. The problems of the present, it is claimed, even if they have been called into existence by technology itself, can be solved and overcome by applying the latest discoveries in the field.

In the past, the so-called technological-scientific method has proved very successful in asserting man's sovereignty over ''inanimate nature.'' The technocrats argue that if we want to progress still further in our mastery over nature, we should apply the same method to areas not specifically technical. The method should be used to analyze man himself, to dissect society, and from there to reconstruct the future. Thus the technology of production can be duplicated in ''organization technology'' and in ''human techniques.''

It is especially the modern planners that are under the spell of this technicistic type of thinking. They promote, sometimes unwittingly, the imperialism of technology and its scientific method. This is particularly clear in the planning procedure they follow.

On the basis of empirical findings, modern planners first draw up universal laws for the way society develops. They look at the past, observe a certain trend there, and then carry this trend into the future. In this way they arrive at certain forecasts for all sorts of areas. The procedure is based on the assumption that knowledge that is useful for explaining the past is equally useful for predicting the future. In cases where it is impossible to speak with absolute certainty, the planners resort to the language of statistical probability.

In a certain sense what the planners are doing is simply extrapolating. Lines of possible (or probable) trends are drawn from the past, through the present, and into the future. To put it differently, a picture of the past is faithfully projected onto a larger screen called the future. Thus the future is a mere projection of the past. Or rather, it is a projection of one's picture of the past. The fact that any picture of the past remains an abstraction of the past is compensated for, presumably, by calling in the aid of multi-disciplinary research and a variety of research methods. Nevertheless, for all that effort, the final step in the whole procedure of the planners is a projection, pure and simple. This is true even when the ''circular model'' is used; that is, past phenomena that revealed a circular or cyclical structure are assumed to repeat themselves as such in the future (e.g., the human life cycle, the seasons, solar and stellar revolutions, and until very recently, economic booms and recessions).

For studying the interrelatedness of the more salient phenomena, such as rising energy demands and diminishing fuel supplies, futurologists resort to modern systems analysis. By means of this type of analysis, they are inclined more and more to try to capture the entire development of our growing world community in one great system with many sub-systems. The activities of the Club of Rome in particular have caught the imagination of many scientists and technologists for a world model of this sort. Simulating systems like these in computers is supposed to provide us with reliable knowledge of what is ahead of us.

These studies by futurologists lead to a second stage: planning or ''assembling'' the future. Plans are drawn up on the basis of the collected data relevant to the future (or to possible futures). The different sub-plans, such as the technical, the economic, the social, and the political plans, are mutually adjusted so that they harmonize, and are then integrated into a single total plan. By using computers to simulate such models of the future, planners can presumably discover whether today's problems will thus be solved and catastrophies averted. The planning phase can be drawn to a close by deciding upon various alternative plans for the future.

In the third phase, a choice is made from among the drawn-up plans, and the plan chosen is then executed. The choice can be left to, say, the economists or to those in the government bureaus, who are the present wielders of power; or it can be left to the people, through the usual parliamentary channels or by direct plebiscites, prepared for by information sent out over the communications media. However, in either case it is a matter of choosing among the plans that are drawn up and worked out by the experts who serve the present ruling powers in society, i.e., by the scientists and the technologists, who will continue to direct the show as the future is being realized, a future that others have opted for but that they have designed. Thus no truly democratic process, direct or indirect, influences the making of the plans, nor is any democratic influence likely to be realized during the phase of executing the plans.[2] The deterministic nature of the plans makes compulsion a real threat, for it will in practice restrict, if not eliminate, the role of free and responsible people. Accordingly, many people recognize that in developing, as well as in executing, such plans we are dealing with a trans-political phenomenon no longer open to discussion. The expertise of an elite serving the present cultural powers is moulding and mastering the future on behalf of all of us.

This picture of a future society does not pose much of a problem for marxist-socialist countries. They have long since merged the economic, technological, and political forces in the power of the State. Besides, Marxists claim that the people's representatives and the party's experts automatically promote the objective interests of the people; this is guaranteed by the very principle of their system of "democratic centralism." Thus, if anything, they reinforce the development towards a totalitarian, collectivist technocracy.

For the "free" west, however, the case is different. If the ideas of the technocrats are to be realized here, the trend of western society will consciously have to be turned, voluntarily or by force, in the direction of a totalitarian technocracy. Naturally, a voluntary transformation is to be preferred and, according to the technocrats, is in any event the most likely to succeed. The criteria they use, after all, aim at an increase in production and, what is more, at sufficient productive employment for all to be able to share in the material fruits of our technological progress. This will assure everyone of survival and well-being. We must therefore simply carry on with the present situation and even reinforce it, and over and above that solve the problems and avert the menaces hanging over us today.

Is this a real future in the sense of a time to come, an open door to new and surprising challenges and opportunities? No, at bottom the technocrats de-futurize the future. Their tomorrow is a day without surprises. They try to bring the future within their grasp by turning it into an extension of the past and the present. Their projection of the future reinforces the present and reaffirms the controlling economic and political powers. On their screen of the future the existing order of things is simply magnified. For this reason these men are also known as the futurologists of order. By reinforcing the technological-scientific order, which is already characteristic of today, they only contribute to the enlargement of the power of science and technology, a power under which people groan or to which they willingly adjust for purely materialistic reasons.

In short, the future of the technocrats, as ideologists of planning, is a technological future, a future in which diverse cultural pursuits and societal relationships are progressively levelled to parts of one gigantic, all-embracing totalitarian system. In that system man will be reduced to a cog in a wheel, to a standardized part of a machine — interchangeable and replaceable. This all-devouring collectivism will culminate in a technocratic world state that will be in complete control of the future.

5

The revolutionary utopians

Among the revolutionary utopians may be numbered such men as Herbert Marcuse, Arthur Waskow, Claus Koch, Robert Jungk, Ernst Bloch, and in a certain sense also Jürgen Habermas. These revolutionary or critical futurologists, as they are also called, are opposed to a rigid, highly deterministic future under the leadership of a technocratic elite. They are opposed, first of all, because in such a future all present misery, suffering, evil, injustice, and oppression will be intensified, not abolished. For instance, even the moral impossibility of a nuclear war becomes a logical possibility in the thinking of a technocrat like Kahn. The reason for this — and this is the second major objection of the revolutionary utopians — is that the technocrats equate history with the advancement and aggrandizement of science and technology, with the result that the present cultural powers are quantitatively strengthened.

The revolutionary utopians oppose the evolution of culture under the leadership of the technocrats by rejecting order and harmony in the development of culture. Instead, they champion protest, conflict, struggle, action — in a word, the *revolution* of culture. For them, revolution is the locomotive of history, propelled by utopian imagination. They bitterly oppose the development of society that leads toward a totalitarian technocracy that gradually imprisons man (even though he will be granted some freedom of movement by the grace of the possibilities of cybernetics). The revolutionaries realize that history and its future will be frustrated if technological progress gains the ascendancy and if work is reduced to purely productive labour, the burdens of which have to be compensated for by the sop of increased consumption. Seeing modern man already weighed down by growing technocracy, they take up his cause, even though he may be so sedated by compulsory production and consumption that he is not even conscious of his plight. The revolutionaries also perceive clearly that the technocrats' solutions to today's problems are solutions that will soon saddle us with even greater problems and threats.

Clearly, the position of the revolutionary utopians is diametrically opposed to that of the ideologists of planning. Rather than defend the inherited situation, the revolutionaries stress the discontinuity of history, the new and surprising elements that may set men free. A pre-determined future is not their ideal. On the contrary, they urge an open attitude toward the future. Yesterday must not be carried into tomorrow; the existing powers must not be strengthened by any increment in knowledge; it is *imagination* that must come to power. In their sketch of utopia, the revolutionaries try to negate the present dehumanizing tendencies and to rough out a future that is open and free. But utopia is not to remain a mere dream. The revolution — especially the political revolution, since it can have the most far-reaching effects — is to be a radical leap out of present reality into the realm of possibilities. In and through the revolution, utopia must abolish itself in the very process of being realized. Utopia thus functions as a catalyst for a radical humanizing process of change.

The formula of the revolutionary utopians, in other words, consists of a categorical negation of the past and the present. The establishment is to be resisted, defied, challenged, and brought low through a conflict culminating in a revolution. Only then will the road lie open to a new future with new possibilities for the coming of the reign of autonomy, the kingdom of freedom, and the rule of peace in what Marcuse calls ''pacified existence.''

They are possibilities and no more than that. Developments in marxist countries have taught the neo-marxist revolutionaries that the revolution does not come with scientific certainty and will not necessarily bring freedom; whoever believes that fails to recognize that the revolution can turn into its opposite, namely, a comprehensive, technocratic dictatorship. Thus warned, they do not plead for a scientifically predictable revolution but for a utopian revolution, which is to be as permanent a revolution as possible. Only in a society in which the revolutionary inspiration replaces acceptance of the existing order and in which revolutionary action takes the place of order, harmony, and technological progress will things be restructured to allow everyone to realize his private and group objectives. The aim should not be to develop science and technology further as instruments of power, but rather to give all possible encouragement to revolutionary creativity and the creative revolution. Over against the logical thought of the technocrats, which they call ''mad rationality,'' the revolutionaries champion a fully rational thought and even more a historical and ethical consciousness. The revolutionary mind addresses itself to the free man, turning against the ideas and thought patterns of vested-interest groups that only magnify the catastrophic present and seal off the future. This revolt is to issue in destruction, for creativity has a chance only in the chaos that comes after the passion of destruction has abated.

Whereas with Marx the revolutionary dialectic ended with the coming of the communist society, with the neo-Marxists the revolutionary dialectic never comes to a standstill. It must not or history would once more become a continuous process, new institutions of oppression would sprout up, and our troubles would begin all over again. Marxism has become questionable to the neo-Marxists because in the transition from private enterprise capitalism to state capitalism the power structures, instead of being conquered, are actually strengthened.

Neo-Marxists therefore espouse a critical dialectic. They are

7

much less confident about the success of the revolution than are the classical Marxists. Marx — at least the older Marx — predicted the success of the revolution with scientific certainty. For the neo-Marxists there is no such certainty. Utopia may well remain utopia, because the establishment will leave nothing undone to restrain the revolutionary forces and prevent them from ever realizing their dream. And it is likely to succeed in this effort if for no other reason than that the revolutionary élan both within and without the technocratic system is still far too weak. The majority of people still fail to see the perils of unbridled industrial technology and still have their eyes closed to the social and political threats connected with more fully exploiting the potentials of the computer and of cybernetics.

The revolutionaries' alternative vision for the future implies a less important role for technology and still has little appeal, because it militates against the current idolization of economic growth, with its promise of higher profits and higher standards of living. The revolutionaries, in particular Marcuse, have been quick to recognize that man has become alienated from nature because of the abstract artificiality of modern technology. They also recognize that man has become alienated from all sorts of things, from his fellowman, and even from himself because of technology's universality, dynamism, and absolutism.

Yet here again it becomes evident that Marx erred. He thought that human alienation could be abolished through communal labour and collective ownership. Nothing could be further from the truth. For while western man has an unprecedented variety of work and jobs to choose from and has little to complain about in the way of material things, alienation has never been so great. Man's self-alienation, it turns out, is more than social and economic. Marx did seek to eliminate the external symptoms of alienation, but he was insufficiently aware of the real needs of man. Moreover he failed to recognize the power of modern technology and the effect it would have on man. Modern man has been reduced to a statistic, he is lonely, he has lost his identity, and he has no real sense of belonging. Today's production-consumption complex conceals only too well that man in truth has become the captive of the constant stream of scientific and technological change. Man no longer has technology under control; technology has gotten out of hand and now controls him.

The method of the revolutionary utopians is no backward-looking method, such as that of the technocrats, who project the past via the present into the future; it is, rather, a forward-looking method, in which the future has to guide and rule the present. Rather than take their starting-point in the science and technology of yesterday and today and thus serve the establishment, they take their cue from a future in which man will be liberated from growing constraint and in which utopia will act as a catalyst for human creativity to set in motion a permanent revolution that will once and for all abolish every form of alienation.

In contrast to the *quantitative* criteria of the technocrats — technological perfection, efficiency, performance, universal order, productive work, consumption, abundance, and progress — the revolutionaries choose what they call the true needs of man: peace, freedom, joy, fun, love, happiness, indĩviduality, simplicity, play, sex. These are needs that come to expression in creative spontaneity and experimentation, in short in *qualitative* changes. They reject the shallow, reductionist image of man current among both capitalists and old-style Marxists — *homo faber*, and *homo economicus*; they choose instead *homo ludens*. Man is not first of all the creature who works, cultivates, creates, and produces, but rather the creature who plays, relaxes, has a good time, and develops himself.

A family quarrel
The revolutionaries claim to offer a future that is open, yet at the same time they voice great uncertainty. And this could hardly be otherwise, for against their better judgement they cherish the illusion that critical destruction of the old order means *ipso facto* the birth of a new and better one. Moreover, their belief that revolutionary changes in society will also bring about a change in man himself and in human aspirations cannot conceal the fact that their passionate yearning for the new order disregards the realities of life. This element, too, contributes to the uncertainty of their future. But the more they realize this themselves, the louder they proclaim that the revolution is almost here. The revolutionary élan in many cases approximates the hurry of desperation.

The major difference between the two categories of thinkers is this: whereas the technocrats want to reduce history to the continuity of technological progress in order to gain control of the historical process, the revolutionaries concentrate history in the present as the decisive moment in which subjective man shakes off the burden of the seemingly objective chain of time and converts the present into a function of the utopia. The present is placed in the light of a new and more human future. Each moment must be charged with creativity and freedom. Only in this way is there hope for a better world tomorrow.

The society of the technocrats, argue the revolutionaries, is an inhuman society of power-hungry men, a society of alienation. In the name of reason, it even entertains nuclear genocide as a possibility, thus insisting that the unthinkable become thinkable and the intolerable become tolerable. The threat of self-annihilation through atomic weapons epitomizes the diseased state of a civilization that

9

prides itself on its unprecedented progress.

The planners, in turn, see the propagated revolution as a clear symptom of decadence, as wanton sabotage of our culture. A revolution, they warn us, will lead to chaos and ultimately to poverty, the very evils science and technology have so magnificently conquered.

There is much that makes sense in the analyses offered by both the technocrats and the revolutionaries and in their criticisms of each other. The important question, however, is from what vantage point we can gain a better insight into the nature of their contest — how it arose and why it developed into the conflict we witness today. In one respect — a crucial one — there is fundamental agreement between the two rivals. At bottom, their quarrel is a family quarrel, a feud between factions of humanism, for neither yields to the other in endorsing from first to last the humanist view of man. Both factions view man as an autonomous being. Both tacitly assume that man is a self-sufficient being who has outgrown the need for God. And in keeping with that assumption, both factions alike take for granted that the world is a closed world and that history is exclusively the affair of man.

But how is it possible then, given this basic unity, for western society to drift into such a crisis? Does man's claim to absolute self-reliance and self-determination entail a civilization divided against itself?

The spiritual roots of the present conflict
We will not really understand the crisis our culture is in unless we first find out how we got into that crisis to begin with. To this end we shall have to go back in the intellectual history of western culture. What is it that has engendered this curious dichotomy in our civilization, in which, on the one hand, we see men with infinite arrogance enlisting the services of science and technology in their cause, but in which, on the other hand, we see men haunted by deep-seated uncertainty, stricken by doubt and despair, and filled, in the face of our present technological culture, with revolutionary desperation?

It is commonly agreed that the development of science and modern technology was possible in western culture only because at the beginning of the modern era man was introduced to a new sense of history and historical development and to a new view of nature and human freedom.[3] Attention was called to *this* world as a world to be cultivated by man. This new outlook, however, lacked internal unity. From the outset two spiritual forces were at work that were moving in fundamentally opposite directions. I am referring, of course, to the Renaissance and the Reformation. These movements gave different answers to the question of the origin of all things.

Basic to the Renaissance was the idea of human autonomy: man discovered himself and proceeded to affirm himself; and before long he came to see himself as the lord of creation. The Reformation likewise saw man as a being called to freedom and mastery, but always in dependence upon God, to whom he remains accountable for all his actions. In Reformation thought, man is less the lord than he is the steward of creation.

At first this basic cleavage in the spiritual foundation of western civilization was not easily detected. For one thing, the key figures often used identical concepts. In the Renaissance, however, these concepts were invested with a meaning that placed man at the pivot of the universe without any transcendental point of orientation. Terms like freedom, responsibility, nature, and control over nature thus acquired a secular meaning. Another reason is that, initially, the practical influence of the Reformation was immense, almost eclipsing that of the Renaissance. Gradually, however, it was the Renaissance view of human autonomy that conquered the world of philosophy and learning. And it was precisely from the intellectual sphere that the idea of human autonomy eventually arrogated to itself the leadership of the whole of western culture.

Hesitantly at the outset but gradually with greater boldness, people denied that man's rule over nature is a gift from the Creator — a concept that necessarily implies a limit to man's self-declared independence and freedom. The western philosopher (and increasingly also the western scientist, under the influence of philosophy) wanted to achieve lordship over nature on his own steam, in his own right, and to his own credit. He tried to base his lordship on the idea of a self-glorifying autonomy, attempting to realize it in his science and to confirm it in technology. And so man's proud place in creation was secularized. Instead of listening to God, he would henceforth listen only to his reason.

Scientific knowledge became the weapon with which man cleared his path toward the future. And this path he could travel with increasing ease as his technological prowess achieved greater heights. People began to believe that man and world could come to self-fulfilment and consummation through the use of modern technology. And so christian eschatology retired from the field in favour of a technological utopia: the hope of a new heaven and a new earth was crowded out by the expectation of a man-made heaven on earth. When, in the nineteenth century, the material fruits of the alliance between science and technology began to mature, the secular belief in progress extended its influence to include the masses. In the meantime, belief in progress has spread so widely that expressions like "the wonders of science" and "the age of technology" have become part of our everyday language.

As the consequences of the philosophy of thinkers like Descartes worked themselves out, man began to occupy the central place in western culture. Nietzsche, and after him Jaspers and Heidegger, has shown how the human ego thus comes to be governed more and more by the will-to-power. Western "egology," which has spawned such terms as self-expression, self-affirmation, self-realization, self-preservation, and self-sufficiency, has found its historic expression in modern technology; its ego-centrism has called up powers that have magnified the tensions in the world beyond imagination.

In the twentieth century, it has become apparent that the attempt to raise the world to a state of technical perfection carries with it enormous drawbacks. The ideal of "peace for all time," which could presumably be attained through modern science and technology, has been rudely shaken by two world wars and continues to be in grave jeopardy. The ideal of unprecedented material prosperity may have been partially realized, yet at the same time it has become clear that gains are often made at the expense of our environment, and that with all our welfare we are sitting on top of a volcano that may be about to erupt.

The belief in progress, with its avowed commitment to unlimited production and consumption, is now threatened by the fact that creation is finite after all, that its resources can indeed be exhausted, and that therefore there is a limit to exploiting it. To add to the grim picture, instead of rendering himself and the world more "real" and more "human", man is discovering that he is actually becoming more and more estranged and alienated from one portion of reality after another. Hence the rise of the neomarxist-tainted revolt against the stifling domination of scientific technology.

If we would better understand the source and the development of the conflict between technocracy and revolution, we will have to look more closely at the compelling lines of thought along which the tension has built up. Both the technocrats and the revolutionaries, as I suggested earlier, stand in the tradition of Cartesian philosophy. Both are guided by the idea of human autonomy. Man wishes to be a god unto himself. The supremely self-confident ego is made the centre of the universe and at the same time the source and origin of all reality.

In this tradition, the technocrats represent the line which declares rational thought to be the immanent origin and meaning-giver of all that is. Man's reason is absolutized on the basis of his claim to autonomous, absolute freedom. Scientific, logical thinking is pried loose, a priori, from its integrated place in life and set apart in a sovereign position of its own. Consequently, the products of rational thinking, too, are set apart in an unassailable realm of their own. The inevitable result of this arbitrary isolation of reason from the total life context is that the absolutized, universally valid mathe-matical and mathematical-physical laws for reality ultimately threaten to subject man's reason itself to an unyielding determinism — and that would surely spell the end of freedom, the very freedom that reason was anchored in at the outset. However, in reaction to this threat, human freedom puts up strong resistance and asserts itself as the counter-pole to rational determinism.

Such, in brief, is the dialectic of modern philosophic thought. And every new attempt at reconciling the polarity is doomed to failure if it is pursued on the same old basis of the autonomy of human thought. When men break the created coherence of reality apart at the outset, by virtue of their pretended autonomy, then so long as they maintain their stance, there is nothing that can put reality back together again.

With the Enlightenment, the dialectical tension sketched above began to burst out of the confinements of mere philosophical theory and to pervade the whole of culture. The people of the Enlightenment aspired not only to understand the world by the light of reason but also to reshape it according to the dictates of reason. Choosing autonomous reason for their instrument, they aimed at developing a society in which freedom could be realized at last.

However, the objective structures, which were first contrived on the basis of reason and then projected into practice, have since turned into powers in their own right, powers which, as autonomous structures, have turned against cultural freedom. And so, on the basis of autonomous, free reason, a technological-scientific society has been constructed. It is a society characterized by virtually autonomous forces that pose a real threat to man's freedom in shaping his culture. And the more dynamically these forces develop themselves, the greater the threat will be, until man is no longer able to oversee the whole, let alone introduce any changes.

It is especially in our day that human freedom is imperiled. Since the time when science opened up the possibility of industrial technology, the new industrial forces have been allied with the political powers, and science and technology have been pressed into service to master the future. Yet people assert their subjective human freedom and resist the technocrats' passion for control. Their resistance is reinforced by the signs that the tension inherent in technocracy — namely, between the desire for infinite expansion and the hard fact of a finite creation — will sooner or later erupt in disasters and catastrophes. The pollution of the environment, the energy ''crisis,'' the risks surrounding nuclear energy, and the growing shortage of raw materials already seem to point in the direction of a collapsing civilization.

For all these reasons, it is understandable that in our day the voices of the revolutionaries find a strong echo. Neo-Marxists in particular speak for those who revolt against the growing technocracy. They turn against the powers of the ''establishment,'' which are used to control history objectively, disregarding the human subject. Their reaction finds its outlet by recognizing man as a free cultural agent and making him the key and cornerstone of all their thought and action. That is why they plead for imagination and creativity and why they work to overthrow the existing order.

Caught in the polar dialectic between absolute determinism and absolute freedom, between the continuum of control and permanent revolution, they choose for revolution and apotheosize man's role in the shaping of culture.

The future

Is the apotheosis of freedom a solution? Do the revolutionaries indeed present a viable alternative to the bleak future of the technocrats? To answer these questions, we must take another look at the nature of the quarrel between technocracy and revolution. The polarity, as I said, derives from a common root: human freedom, declared autonomous and made absolute. It is this root that first generates the passion for control, a passion that then pushes us in the direction of technocracy. Soon technocracy begins to lead a life of its own, threatening to obliterate human freedom. Men thereupon take up positions at the freedom pole and declare war upon all controls. Obviously, the two poles simultaneously presuppose, penetrate, and repulse one another.

As a result of this mutual penetration and repulsion, the technocrats continually run into the "irrational factor" called human freedom, in spite of all their efforts to exorcise it. The progress of technology, of which they boast, is only possible if human beings continue to be free to use their ingenuity and inventiveness; the very extension of technocracy is only possible if human beings continue to make decisions in its favour.

Meanwhile we also observe the opposite: the revolutionaries, for all their desire to be absolutely free, can in fact never leap free of continuity, control, or power. As a historical or cultural agent, man is bound to objective cultural means and cultural power. Without these he cannot express himself culturally. It is for this reason, too, that we often see revolutionaries either going over to the camp of the "establishment" or making their revolutionary programmes still more radical. In the latter case, they may either promote total chaos or make a lunge for power themselves. Either way, the revolutionary ideals are given up, the revolution's own children are consumed, and a dictatorship sets in that is more powerful and despotic than the preceding one. From this point on, the cultural dialectic from establishment to establishment and from revolution to revolution can only increase in magnitude and intensity.

In this conflict over the direction of culture, the technocrats appear at a decided advantage over the revolutionaries because they need not rely on men as free historical agents. Instead, by sheer economic power they can capitalize on the objective cultural potential as it becomes available in the most recent scientific and technological possibilities, such as systems theory, cybernetics, and the computer. The technocrats enjoy an additional advantage because the masses either depend so completely on them that they are impotent or they surrender body and soul to them in the confident hope of receiving still more of the good gifts of science and technology.

Moreover, when the technocrats are confronted with embarrassing problems or imminent perils, they simply respond by changing their strategy. And as they do so, they do not spare human freedom; in fact, if need be they restrict it even further. This tendency has been evident, for example, at recent world conferences devoted to such pressing problems as the population explosion, environmental pollution, and the economic development of third world countries. In the face of all the clamour for more rigid controls, it should come as no surprise that the revolutionaries become even more radical; to achieve their goal of overthrowing the existing order, they may resort to greater instruments of destruction.

If my analysis is correct, the tension in our culture can only grow to more disastrous proportions. Technocracy is becoming increasingly centralized, and it threatens to encompass the entire globe; its tight control will turn our world into one great prison-house for free and responsible people. Revolution, on the other hand, if consistently realized, will inevitably result in violence and destruction. Therefore, if a third way is not found, the only choice open to mankind is between a technologically streamlined society of perfected standardization and mass culture and a cultural self-annihilation and suicide.

There is a real danger that science and technology *as such* will be blamed for our present dilemma. In many quarters, in fact, people have already come to this conclusion. But then the nature of our crisis has been woefully misunderstood. It is not science or technology but *man* that bears the blame. Western man has chosen to accept this world and himself as his first and his last point of reference. He has gradually closed his eyes to any transcendent reality. The purpose of history and the meaning of life have been restricted to this world; they have been made immanent. And man, no longer open to God, is now thrown back upon a purely this-worldly reality.

All the same, the western mind suffers from the fact that divine revelation once instilled into it notions of perfection and consummation, notions that retain their appeal and that refuse to be silenced. However, since western man no longer looks to God for the fulfilment of these promises, he is obliged to arrogate to himself the task of realizing them. Thus, as he moves further and further away from God, man secularizes God's promises and begins to think that he can realize these for himself through science and technology. Having placed his faith and confidence in technological progress, he appears to have thrown himself and his future at the feet of technological-scientific development.

In this light, one can well understand that where true history is fundamentally closed off and a purely man-made history is advocated in its place, modern technology in its many applications must grow to inordinate dimensions. In fact, it is already assuming monstrous proportions and is beginning to betray features that are actually demonic.

As the process of secularization widens and deepens, as man's sense of responsibility diminishes, as his hopes are increasingly pinned on science and technology, and as the possibilities of the latter become more frightful, the thinking of technocrats and revolutionaries actually approaches two rival forms of nihilism. The house of the closed world-view is a house divided against itself. It is the natural home of cultural tensions and catastrophies. The rule of technocratic control is self-willed, man-made, power hungry, and hence normless. The reaction it elicits is an equally lawless revolutionary freedom, in which every last cultural achievement must be annihilated again. The nihilism of the lifeless mechanical order of the technocrats has its obverse in the nihilism of revolutionary turmoil and chaos.

Is there hope for the future? Unlimited technological-scientific development will lead to loss of freedom, the exhaustion of nature, and possibly the destruction of the world. The unleashed revolution, intended to liberate man, will only lead to greater slavery. This nihilistic dialectic, which is growing in scope and intensity as history moves forward, marks the way a culture advances toward its dissolution.[4]

Estranged from God, our civilization carries about within itself the seeds of its own death. Already it is experiencing the blight of decay, which will not stop spreading until all lies in ruins. In all of this, however, God himself is clearly making us feel that apart from him there is neither life nor survival.

The way out
As I indicated at the beginning, I have restricted my analysis to the extremes of the opposing tendencies that divide our culture. Fortunately there are still sufficient counterforces to keep the hopeless dialectic from breaking our society apart. However, no counterforce can change the fact that the two tendencies are working in opposite directions and are causing serious cracks in the

foundations of our society. There is not a person who does not experience daily something of the increasing tension between technocracy and revolution. We are left with the pressing question: is there a way out?

In order to point a way out of the predicament, I must first focus on the problem of human autonomy and the attending secularization of culture, for it is autonomy — or rather the pretension to autonomy together with the closed world-view of the secular mind — that constitutes *the* problem of our time.

For example, man's autonomy cannot possibly be absolute. Man is and remains, in technology as elsewhere, fully dependent upon *given* materials and structures. And, likewise, he is and remains dependent upon himself, in the sense that he cannot ground his own existence in his *belief* in autonomous, independent existence. His very mortality shows up the emptiness of his claim to being his own origin, his own source of life, and his own god. Where people do indeed perceive this much, however, they often end up resigning themselves to human existence as to being-unto-death. But the realization that death is the end then infuses the whole of life with meaninglessness. The closed world view of secularism ultimately allows no other vista than that of nihilism: a prospect without hope.

A different perspective will require a different mentality. Man must recognize that he is incapable of autonomously determining the direction in which culture should develop. He must wake up to the fact that he cannot pretend to be the lamp that lights the road he should go. So long as he keeps up this pretension, his world — as history since the days of the Enlightenment clearly shows — will continue to grow darker and more menacing all around him.

I said earlier that modern man labours under alienation from self, from his fellowman, from nature, from culture, and from history. The deepest cause of this ailment is that man has become alienated from the origin of all things, God. Alienation from God always brings with it the other forms of alienation and must finally issue in the utter meaninglessness of everything.

Anyone who has seen and acknowledged the deepest cause of our cultural crisis knows that there is a better way. It is the way in which man is not the measure of all things. It is the way in which man is conscious of being carried and guided by a Creator God, the God who has given him life and who has crowned him with honour and dominion for the sake of responsible stewardship. It is the way which requires that man be open to the meaning of history and to the meaning of his historical existence. The meaning of human life on earth cannot be found in visible, temporal reality itself, nor in any part of it. To restrict meaning to created reality is to constrict it. It is to close meaning off and ultimately to choke it by isolating created reality from its life-giving Origin and by deifying the creature.

17

Meaning transcends reality. When we recognize that, a horizon opens up that stretches beyond the horizon of this world, with its tensions and its distress, with its sin and evil and death. Before that opened horizon, life has a future again.

I would emphasize that this way is shown us not by philosophy but by God himself. We know from the divine revelation of Scripture that God in Jesus Christ is Lord of history, that he alone rules and governs the world, that he holds all things together and brings them to fulfilment in the final consummation. With Christ's coming into history, the kingdom of God has broken into the world of man, to conquer and to heal it. Consequently, the quest for the kingdom must also be expressed through responsible activity in science and technology.

Only after confessing belief in God and his sovereign rule over the whole of human life can we find a philosophy that orientates itself to that rule and kingdom and that can be of service in indicating the way of deliverance. From that point on, a christian philosophy of culture and a christian philosophy of technology have their work cut out for them. For they must point out solutions and avenues of escape, especially at those junctures where the development of modern culture has tied itself into knots.

Some implications
The implications of what I have been saying so far are many. These cannot all be worked out here, since my chief concern has been to make a proper diagnosis. Still, I should like to examine briefly what a christian philosophy can say of significance about science, planning, and the meaning of technology.

(i) Science
The predicament of our "scientized" culture calls for a re-evaluation of science. It is especially important that we critically re-examine the absoluteness with which scientific "truth" is presented. First, people need to recognize that "science" as such does not exist. Only flesh-and-blood scientists exist, and they hold certain scientific theories. These theories are not, as is so often believed, objective and neutral; rather, they rest on believed assumptions and are based on hypotheses. This is the reason scientific theories are always *conditioned*, coloured theories. Further, scientific theories are *relative*, relating as they do only to the knowledge of certain aspects of reality, such as the physical, the biotic, the social, the economic, among others. Therefore, scientific knowledge is necessarily a knowledge of limited scope, abstracted from the fullness of reality, which is itself far more complex. In addition, since reality's complexity even penetrates each of its aspects, scientific knowledge as knowledge of a certain aspect must always

be a knowledge for the moment, never finished, never complete; in short, it is *limited* knowledge.

The view that scientific knowledge is conditioned, relative, and limited was never more needed than it is today. For example, more controls on people's activities are inevitable because of the uncritical use of scientific knowledge in systems theories and quantification and the facile recourse to computers in conjunction with these methods. To conduct science naively and to apply it indiscriminately is to cast out human responsibility. If scientists admit the qualifications and limitations of science, however, the knowledge that science yields is taken up into a more comprehensive, responsible knowing. Instead of making human responsibility unneccessary, the growth of scientific knowledge heightens the need for it. Moreover mankind's heightened responsibility is becoming increasingly a *communal* responsibility, and this development requires a much stronger sense of community among people than has been experienced up until now. It is a condition that can really be met only if there is a common sensitivity to norms and a common religious consciousness. But as everyone knows, these are the very things our secular culture lacks most.

(ii) Planning

The method so frequently applied in planning the future is the so-called technological-scientific method, the same method that is used in manufacturing lathes, computers, wireless receivers, and so on. Using this method in planning for the future entails the grave danger that people are going to be manipulated as though they were mere things or simple parts of a machine. Planning is intended to banish evil, but its practical effect is to abolish man. Planning cancels out man as a free and responsible creature made in the image and likeness of God. If the finished plans are ever carried out, man will increasingly be enslaved to powers beyond his control.

This outcome is so certain because throughout the phases of planning — doing the research, drafting the plans, executing the plans — it is science and the scientific method that men allow to have the first word and the final say. In the first stage of research, moreover, scientists often simply accept the facts at face value without questioning their normativity. In the second phase, too, this question is rarely raised. Also, the planners are presumptuous as they ''assemble a model for the future''; they forget that it is not given to mortal man to draw a picture of the future that takes into account all the factors that will then be at work. What the planners lack is enough humility to halt before the unknown, before the unknowable, unexpected, and surprising aspects of the future.[5] The adverse effects of their presumption are reinforced twice over when the planners integrate their various sub-plans into the total plan:

19

man is levelled, society is collectivized, and power is concentrated in the hands of a few. In the third phase, accordingly, when the plans are being executed, there is little if any room left for man; he can no longer play a free and responsible role in shaping the world of tomorrow.

Most people would agree that our society has become so complex that it is no longer possible to make intelligent decisions about the future without using scientific knowledge and resorting to scientific analysis. But science must not lay down the law for man or dictate to him what to do. It should elevate his personal and communal responsibility, not eviscerate it. When science is thus subordinated to human responsibility, men will also be free to explore any new developments that could open up unexpected opportunities for solving problems that now seem insoluble. Today's planners often force their way to a solution in a high-handed, technocratic manner, blocking the road to genuine solutions.

As a rule, planning involves integration and collectivization. This tendency, too, must be resisted. If we wish to give priority to human responsibility, we need to provide for differentiation alongside the tendency towards integration. Where so many things are done on ever larger scales, there ought to be more room for activities organized on smaller scales. Promoting only the super-size ventures leads to power concentrations, collectivism, and drab uniformity. Smaller, differentiated ventures not only create excitingly new opportunities for human involvement and inventiveness, but also help develop a pluriform culture by spreading authority and competence, in both the economic and the political spheres, over a greater number of people and a greater variety of institutions.

One other matter should be clear from my analysis: the many different human communities and relationships must not be denatured into subordinate parts of a technological world state. Rather, they should be respected according to their own normative structures.

The cardinal question in any human relationship is where the line between authority and freedom lies. To decide that question requires much wisdom and reflection. But whatever the answer, two evils must be avoided. On the one hand we must avoid any absolute authority of persons and rule books, of anonymous powers and the exigencies of technocracy. On the other hand, we must avoid unbridled revolutionary individualism.

Within any given relationship, be it a family or a school, a business or an engineering firm, the normative structure for leadership on the one side and subordination on the other is meant to serve a healthy unfolding of life. When the normative structure is not acknowledged, it will inevitably assert itself in a corrupted fashion, either as dictatorship or as anarchy. In either case it will then be impossible for the members of the relationship to serve each other in love — which is the very purpose of the normative structure for leadership and subordination. Those who are in authority and must give leadership have a responsibility to create conditions that will enable everyone to work at his task in optimum freedom; they ought also to prevent any disturbance of the community by intrusions from outside or by internal disruption caused when members misuse their freedom. Those who are in a subordinate position, meanwhile, have a responsibility to work together to achieve the purpose of the relationship. To underscore the communal task and to forestall misguided or biased decisions, members need to help, encourage, and correct one another. Fruitful interaction of this sort is realized best if policy and goals are discussed in an open atmosphere and if those who carry out policy are required to give an account of their doings at regular intervals.

I admit that these guidelines are anything but easy to work out in practice. Nevertheless, they offer a sound framework for forming genuine communities of men.[6]

(iii) The meaning of technology
The meaningful functions of technology are many. They include emancipating the body and the mind from toil and from drudgery, repelling the onslaughts of nature, providing for man's material needs, and conquering diseases. They also include eliminating unnecessary burdens, freeing time, promoting rest and peace, and evolving new ways and means for advancing the disclosure of culture. The meaning of technology concerns the elevation of culture — through fostering reflection, through stimulating internal communication, and through making possible a wide variety of rewarding jobs and tasks.

All this is a far cry from what technology actually is today, for the meaning of technology has been perverted. Partly under the influence of economic power structures, technology has produced superfluity, waste, and pollution. Work has been reduced to ''productive'' work, to work that is ''economically justifiable''; and the resulting emptiness in the workers is compensated for by more consumption. Instead of freeing men to devote themselves to works of assistance and service, of care and mercy, of creativity and beauty, technology has all but banished these forms of meaningful labour.

This impoverishment, as I noted earlier, developed because western man began to *believe* that technological know-how, assisted by economic efficiency, would bring cultural progress. Technology was expected to deliver what it never could: the redemption of life. People produced whatever could be produced as materialism, pragmatism, and positivism reinforced their faith in a technology that was absolute and intrinsically anormative. Technology has buried life under the yoke of a terrifying power.

If we want to experience the true meaning of technology again, we will have to abandon our mad pursuit of the future, obsessed as we are by our technological prowess, by our will to power, and by our mania for consumption. We will have to take up the battle against superfluity and absurd luxury. To answer to the meaning of technology we will have to devote all our powers in love to the development of our fellowman, of our natural environment, and of ourselves, in happy accordance with the ultimate purpose of each. This focus will guide us is making conscious, responsible choices about *what* will be produced, instead of being governed in such choices by our insatiable thirst for continuously expanding our technological development. The autonomous dynamics of technology will have to be decelerated, so that we may have time once again to reflect upon the meaning of it all.

We will have to determine how much of the damage we have done can be set right. Perhaps we need to recover some forgotten traditions in the history of technology and adjust these to our needs. The so-called alternative technologies will have to be given more than the usual attention; I think, for instance, of durable technical products that can be made with little capital investment and low energy expenditure. That approach would involve using as much as possible the natural materials and sources of energy available to us. It would result, moreover, in individual products that give satisfaction to those who make them, that are geared to specific cultural patterns, and that cause a minimum of pollution. Such scaled-down forms of technology seem almost unreal in the overpowering presence of our gigantic forms; indeed, they are not easy to realize, and they will certainly make heavy demands on our technological imagination.

Alternative technologies offer one way to attack the problems that present technology has fostered in the workplace: the rift between work and leisure and a growing separation between the worker's living quarters and place of work. These obvious problems could be reduced by developing new trades and crafts that use modern technological equipment and also by using modern information and communication techniques on a larger scale to include private users.

In short, to probe the meaning of technology does not mean to throw technology aside wherever possible. Rather, it means to appreciate technology's proper and meaningful place within culture and to develop technology intensively and responsibly.

We must not allow technological-scientific possibilities and economic forces to dominate our culture. Rather, spiritual and cultural values must make technology serviceable to life. That means technology must be opened up socially, so that all who are involved in it are entrusted with responsibilities. Disclosing technology implies, further, that the economic development of technology should not be limited to maximizing profits or catering to consumers; we should prevent waste and strive for a frugal use of nature, even though that will inevitably mean cutting back "economic growth" and tempering consumption. In addition, when we eliminate harmful waste products and when we refuse to use nature as our rubbish heap, we are showing that we have an eye for the aesthetic dimension of technological development; technology must not mar nature but rather be developed in harmony with it. Further, justice demands that we preserve nature, not imperil it, by keeping it clean. Where nature is being destroyed, our laws and courts ought to intervene and punish offenders, both to protect nature against mutilation and pollution and to protect people from the dangers inherent in a deteriorating environment.

The disruptions we are experiencing in our culture only arise when the ethos of men is wrongly directed, so that in all their technological doings they myopically focus on something within created reality, absolutizing and thus asphyxiating it, instead of offering themselves and all their deeds as living sacrifices in the service of God, who in Christ rules over creation. Man needs to be delivered from his shortsightedness and from his satisfaction with short-run measures of relative success. Only the wider perspective of the kingdom of God can accomplish that. Only through openness to this transcendent reality can western culture look forward once again to a meaningful future.

If we try to implement these guidelines in our practical affairs, we will find that what stretches before us is not a dead-end road but a highway of liberation. This highway alone avoids the stagnation and the entropy of technocratic culture. This highway avoids the kind of development that kills human initiative and that consolidates an elite in citadels of collectivized and centralized power. This highway, too, keeps our culture from being pulled into the vortex of the ideology of revolution.

Notes

1.Another recent current in western culture that I shall not discuss in this essay is the passive counterculture. This movement suffered from much internal disagreement, divided as it was over a number of short-lived, competing subcultures (cf. Alvin Toffler, **Future Shock**, chap. 1). Together they formed a persistent and intriguing phenomenon on the fringe of western civilization. Reacting against the lack of freedom of technocratic society — in this respect they agreed with the revolutionary utopians — the members' of the counterculture chose for the impossible: a culture with bare hands and on bare feet (cf. Theodore Roszak, **The Making of a Counter Culture**, pp. xi, 50 ff). Worse still, they had a tendency to switch at any time to the camp of the revolutionaries (ibid., pp. 61,63). Their terror of technocracy was expressed by their escape from culture into romanticism, pantheism, and naturalism — as evidenced by their fascination with hallucinogenic drugs, their enthusiasm for astrology and occultism, their frantic search for truth in the expanded consciousness and the ecstatic experience, and their radical rejection of science and technology (their continued dependence on them notwithstanding). For a critical discussion, see Os Guinness, **The Dust of Death** (Downers Grove, Illinois: InterVarsity Press, 1973).

2. Direct democracy, as it is called, even facilitates research of the future since one can make use of the laws of large numbers to reduce the surprise element in the future by basing one's forecasts on the verdicts of the people. I have worked this out further in my forthcoming book, **Technology and the Future.**

3. Cf. Herman Dooyeweerd, **Reconstruction and Reformation,** chap. vii. See also R. Hooykaas, **Religion and the Rise of Modern Science**, esp. pp. xiff, 98ff.

4. This nihilistic dialectic is apparent in Theodor Wiesengrund-Adorno's **Negative Dialektik** (Frankfort, 1966), p. 338: "One should try to live so that one may believe one has been a good animal." Cf. Günter Rohrmoser, **Das Elend der kritischen Theorie** (Freiburg, 1970), pp. 31, 34: "According to the school of negative dialectics, in a world impounded in its own inverted order the extent to which one can still realize something of his humanity is to be a good animal." For this reason it is more appropriate, in my opinion, to speak of **nihilistic** rather than **negative** dialectics.

5. In this connection, compare Deuteronomy 29:29 and Revelation 10:4.

6. Cf. Hendrik van Riessen, **The Society of the Future,** pp. 290 ff.

The environmental problem: its neglected religious-philosophical backgrounds

Environmental pollution is a subject widely discussed today. Daily the media confront us with the looming disaster of a polluted, deteriorated environment in which life's suffocates and man's outlook for the future is hopeless. Higher temperatures, oxygen consumption outstripping its production, increasing carbon dioxide in the atmosphere, menacing radioactivity, poisoning of soil, water, and air — together these symptoms are the ingredients of a very pessimistic prognosis for the survival of mankind.

The report of the Club of Rome, however, pointed out that there are more factors than these involved in the ruinous course our world has taken. These factors, such as the diminishing supply of natural resources and energy, the population increase, and the concomitant problem of food supply, have a bearing on the environmental problem. Nevertheless, I am going to focus our attention particularly on the problem of pollution itself. And even then, I must restrict myself to one side of pollution.

The restriction is indicated by the subtitle of this essay: the religious-philosophical backgrounds of the problem. This may be surprising because it is far from customary to associate man's religion with the pollution of the environment. Usually this problem is approached from the standpoint of a special science, such as biology, technology, or economics. Even the answers proposed are generally limited to scientific or technological ones. Yet I think such approaches tend to touch only the surface of the problem, both in the discussions and in the solutions offered. Faced as we are with the enormous drawbacks that have attended the development of science and technology (such as the destruction of the environment), it is essential that we not flee from the crisis in order to seek refuge with science and technology themselves. Instead, we must enter into the religious heart and source of our scientific activities. For it is in this religious background of our technologized culture that we shall discover the origins of environmental pollution.

This kind of investigation does not occur very frequently because we are too impatient to await the results of new reflections on actual problems. What's more, the scientific and technological enterprise has occupied us to such a degree that the deeper currents beneath the problems generally do not receive any attention. Man no longer seems able to perceive that science and technology are based on philosophical views and that these views, in turn, are rooted in religious convictions. However our blindness does not mean that these religious-philosophical influences do not exist.

Man expresses his fundamental convictions concerning reality and its questions in his religion. This basic religious attitude is the foundation for his living and hoping. It also inspires his philosophical thinking. Philosophical thought is not neutral, objective, or devoid of any values, but is inspired religiously. This religiously inspired philosophy, in turn, exerts its influence on the special sciences, which are related to that philosphy. Furthermore, modern technology, based as it is on modern science, especially modern natural science, also experiences the influence of religiously inspired philosophy.

Therefore, philosophy, inspired by basic religious convictions, plays a decisive role in the development of science and technology. The religion of western man leaves its sediment in philosophy, and this religiously inspired philosophy influences the views of the special sciences. These views, in turn, assume a tangible form, as it were, because they are projected into daily life through their application in the development of modern technology.

Thus, in the face of the massive problems and inevitable calamities brought on by modern technology, we feel driven to investigate the content of the religion operating beneath the surface. Considering the enormous technological dislocations in the human environment, we must ask whether or not the error ought to be located especially at the spiritual level.

If my interpretation is correct, the causes of pollution lie in the religious origin of western science and technology. Therefore, we risk giving myopic and irresponsible guidance if we concentrate all our attention on the startling and, at times, overwhelming character of pollution. For we need to realize that what we see surfacing on a large scale today has determined the content of spiritual attitudes for centuries. Long contained by man's view of nature and of himself, it is now violently emerging. Only when we discern the spiritual-historical background of the environmental problem will it be possible to go beyond a simple appeal to science and technology to avert the imminent dangers to an appeal that involves modern man's religious convictions, which affect his attitude towards nature and culture.

God, man, and nature

It is difficult to understand my analysis unless we go back in the intellectual history of mankind. Initially, man lived in complete harmony with nature in the Garden of Eden. He did not need to plow or sow, for trees, "planted" by God himself, produced fruits and sustained him. This harmonious relationship between man and nature was changed drastically, however, by man's fall, in which he rejected God and wished to be a god himself. From that moment on, man had to earn his bread by hard work; he was constantly threatened by his natural environment, all the more intensely as his apostasy continued. In fact, men came to believe that nature was determined by mysterious powers. They conformed to the demands of nature and did not often have the courage to intervene. But when they did, they also carried out magical rites to propitiate the gods of nature.

It is true that a few changes took place in the ancient Greek world, but none of these was fundamental. Reality was seen as one organic unity in which everything had a fixed position. The ancient Greeks thought of man as inserted into a hostile nature. Its capriciousness could be averted only by concentrating on the stability and immutability of the supernatural, the world of the stars.

During the Middle Ages, men came in contact with the divine Word-revelation and realized that nature is not divine but is instead a created reality. Nevertheless, the predominant view of nature was similar to the Greeks' view. Greek ontology as a philosophy which deals with the totality of beings was simply adapted to fit the understood meaning of christian revelation. Men were not yet motivated to examine the various functions of nature, since the world view of the day contended that everything, including nature, had its fixed and immutable place. Because nature formed a given and static unity that did not tolerate any intervention, men resisted a dynamic development of science and technology. They feared changes because they could disrupt and disturb what had been given to man. This cautious approach to nature also characterized the development of the crafts.

Basically, the medieval attitude toward nature rested on a dualistic view of reality. A firm faith in the hereafter for the soul prevented people from paying adequate attention to this world and to nature. As a result, man's investigation of nature did not go beyond a hierarchical ordering of natural data. Nature was interpreted in terms of what was alive. This organistic view of nature implied that whatever in nature was judged inorganic or "dead" had to be regarded as less than alive, and therefore not worthy of full attention.

At the beginning of the modern age, this situation changed. During both the Renaissance and the Reformation, people began to pay attention to this world on the grounds that it needed cultivation. The people of the Reformation did not regard nature as divine, and they denied the existence of a divine and static order for nature that man had to leave untouched. While they recognized the legitimacy of science and technology, they emphasized that science and technology had to be normed by God's Law, which held for the whole of created reality. The people of the Reformation could not accept science and technology operating autonomously, independent of the creator of all things. Instead, they said scientific and technological activity were performed in obedience and praise to God rather than to man.

A starkly contrasting view emerged from the religious attitude of the Renaissance and the later humanism that dominated the development of science and technology. During the Renaissance man was proclaimed completely autonomous, independent, and self-reliant. Aided by science and later by modern scientific technology, man attempted to make his life secure by developing science and technology in absolute independence. Initially, the influences of the Reformation, the Renaissance, and the later humanism were largely the same. But since the Enlightenment, the development of science has been inspired chiefly by the religion of humanism, which regards man as sufficient in himself. That does not imply, however, that Christians no longer participated in developing science and technology, only that their christian religion became increasingly isolated from their scientific and technological endeavours. They seemed to hold a dualistic view of reality and human activity; their christian view was adapted to the humanistic tradition, while the authentic character of the christian religion disappeared.

For a proper understanding of the religious backgrounds of the problems surrounding the environment, therefore, we must examine the tenets of the humanistic religion, particularly its view of nature. Cartesian philosophy illustrates these tenets quite well. Descartes doubted all tradition and belief. His search for religious certainty was limited to man himself; he found it in scientific thought, which was considered to be the ultimate certainty and truth. Descartes's philosophy was determined primarily by two substances: *res cognitans* and *res extensa*, thinking man and space. They were treated as two opposite poles within reality, but thinking man took the primary place. That is to say, everything that exists beyond and outside of man was interpreted from the vantage point of thinking man.

While Descartes reached this conclusion with the help of mathematics, Galileo reduced everything to the object of natural science. The earlier stress on the being of beings was replaced by an emphasis on the function of beings. Previously, men had regarded God as the origin and therefore also as the unity of all beings, but this unity was increasingly bestowed on thinking man. The human subject became the central point of reference for everything that exists. In modern philosophy, man became the predominant centre of the world. The human subject became pivotal, while nature was relegated to the status of a mere object of natural science, which dealt with reality in terms of cause and effect.

As a result, nature was interpreted mechanistically. It was looked on as a whole made up of interacting forces that could be calculated and expressed in formulas. In fact, reality was thought to be an ingenious mechanism. Animate and inanimate nature were related to this mechanism in degrees of complexity.

From its outset, this mode of thinking housed the possibility of reducing nature to a mere object of technological manipulation. Man was no longer seen as a child of Mother Earth or as part of the whole of natural reality. On the contrary, henceforward man was to be the autonomous lord and master over nature, a power he had to demonstrate. This meant for technology that man was no longer limited to using the things nature offered and that he was able to take from nature as he saw fit. The so-called organistic view of nature was replaced by the mechanistic view, in which man, aided by technology, shaped nature to match his wishes. The natural process was pictured as inanimate but ordered, and its irrevocable and binding laws could be revealed by mathematically inspired natural science. The greatest danger inherent in this view is the absolutization of the method that it gave rise to. The natural-scientific method was thought to be superior to all other scientific methods, and everything that did not come within the scope of this method was written off as unreal, hence negligible. The abstract knowledge peculiar to natural science was thought to be a full, concrete knowledge of reality.

Therefore in humanism man no longer saw nature as created and upheld by God. On the contrary, he denied that nature is intrinsically related to God and has a sacred character. Moreover, humanism rejected the existence of a divine order of things to which nature is subjected and which does not tolerate human interference. I must emphasize that I only partially agree with the second interpretation. For we need to remind ourselves continually that the humanist view was completely secularized. It was a view that regarded nature as a mere workshop at the arbitrary disposal of sovereign and autonomous man, the pretended creator of nature, who could shape the order of nature as he pleased. Of course, this absolutely independent sway over nature was packaged with many promises: for instance, that there was no need to continue living under the threat of a capricious and hostile environment. Instead, doors were opened to a world that was *man's*, a world that he designed, controlled, and subjected to himself.

This, in short, was the radical shift brought about in man's relationship to nature. Modern natural science removed the ethical and religious restraints of earlier times, hesitantly at first but with increasing boldness as confidence and momentum grew. Those views of reality that did not concur with the dictates of natural science lost ground until they were denied altogether and finally vanished.

As a result, only one attitude towards man and his relationship to his surroundings appeared to be valid and worthy of respect; this, of course, was the one prescribed by modern natural science and its method of quantification. This method was used to interpret everything exclusively in terms of numbers and quantifiability. That is to say, everything was *reduced* to numerability, measurability, and estimability and was thus prepared for calculation and subsequent technological control.

The mechanistic view of nature monopolized man's thinking and changed nature's rich diversity into a frightful, monotonous homogeneity. As measurement, estimation, and enumeration became the only valid ways of approaching nature, its inherent coherence disintegrated.

Initially, this novel attitude remained restricted to the area of science. Its practical consequences, therefore, could hardly be detected. At the beginning of the eighteenth century, however, this situation changed because of the advance of modern technology based on modern science. The economic and material circumstances of the day also contributed to the activity of modern technology, while the population increases so stimulated its further development that it seemed imperative. Fueled by the additional available manpower, only the development of modern technology could satisfy the tremendous growth in human needs.

Modern technology did not develop on its own steam. For the spirit of the Renaissance and of humanism, which involved a firm faith in the progress promised by technological development, was also at work. Previously, this faith had been embraced by scientists only, but now the masses accepted it as well; thus its influence was extended to everyone who did not object to the new and infectious prospects of riches and liberty instead of obligatory poverty and suppression. Modern technology became a liberating force.

It is very important, therefore, that we keep sight of both the influence of natural-scientific thought on modern technology and the absolutization of this thought, which inescapably leads to the objectification and constriction of the meaning of nature. For men interpret nature primarily in terms of mathematical categories, such as the numerical aspect of space and the physical categories of time and causality, under the illusion that nature is composed entirely of these primary characteristics; meanwhile, others, such as colour, smell, and sound, are under-valued. To be sure, these characteristics are still called secondary qualities at first, but gradually they are ignored altogether. Inevitably, man misjudges the intrinsic value and meaning of nature because he no longer recognizes nature's essential and normative meaning. This attitude is illustrated in the works of Karl Marx, for instance, when he wrote in his Paris manuscripts that "nature, when seen abstractly and in itself, has no meaning for man when alienated from man."

We should not conclude from this that objectifying and constricting the meaning of nature brought about immediate large-scale dislocations the moment modern technology began to develop. Although various people pointed out the inevitable disastrous effects of allowing technology to dominate human life without any resistance at all, such warnings were not heeded. And as a result, nature today, strangled as it is through the advances of technology, harbours potential disasters. Instead of enabling man to live, nature has ceased to fully support life. Mankind's consistent application of the mechanistic view of nature produced the threatening results that confront us even now.

The application of natural-scientific knowledge was not a matter of coincidence. Descartes had already pointed out the tremendous usefulness of the new view for meeting man's needs. With an expanding knowledge at his disposal, man would master all forces and movements in nature until they conformed to his wishes and worked to his advantage, simply because man was the possessor of nature.

Thus nature functioned as the object of human knowledge, while at the same time it served as the object of human mastery, satisfying man's arbitrary cravings for utility. This is clearly demonstrated by a statement from Marx's *Outline of a Critique of Political Economy*: "The material of nature, insofar as it has remained untouched by human labour, has no value, for value is but substantiated labour...."

Since the time of Descartes and Galileo, therefore, the modern scientist can be described as a rational engineer, who autonomously directs and controls the forces and movements inherent in nature, rendering them useful as he sees fit, for he is responsible to himself only. In other words, the compulsion to dominate the world by means of technology precedes modern technology itself.

Francis Bacon was particularly appreciative of the potentially positive effects of the new view of nature. He anticipated the achievements of today's technology, but he failed to pay sufficient attention to its possible negative aspects. Bacon combined the persistent desire to apply natural-scientific knowledge with the idea of unprecedented material progress, making him the first to believe in progress brought about by means of science and technology. According to him, man had the duty to pursue new scientific knowledge to establish his unprecedented power.

31

The dominant view in modern technology, therefore, is that man is in a position to command the world as he wishes. With technology as his tool, man sets out to create a world in which he alone is lord and master. He is motivated and stimulated to do this because of his need to safeguard his autonomous position, so that he may continue to enjoy and consume the fruits of his own labour.

Such ambition will result in a fixed and constricted technological world, far removed from the rich fullness of the whole of reality. The man-made technological world is but a fragment of the whole world and reality. Yet people persistently view the fullness of the whole of reality as if it coincided with the world of technology. In other words, modern man has blown up his desire to dominate by means of technology until it fills all of reality.

In so doing, man reduces reality to one of its aspects. Reality, however, will not tolerate such reduction, because it consists of a diversity of aspects. Everything in reality exists in a coherence of meaning given with creation itself; man cannot reject this coherence without suffering the consequences. Concealing this coherence or failing to reckon with it will inescapably result in serious derailments and dislocations. As technological control becomes absolute, ruinous side-effects will begin to appear. To be sure, this tendency can be ignored for a short time. But with industry growing, with agriculture almost completely industrialized, and with traffic increasing continually, this trend will assume such proportions that it is safe to speak of the qualitative doubling of disruptive side-effects. People will be unable to ignore such a situation for very long. It is these side-effects that together constitute the problem of environmental pollution today. The nature that we control and dominate threatens to turn on us. Destroyed and polluted, it has become a definite threat to the survival of mankind. The religious faith in progress has combined with technological progress itself to bring mankind to a critical stage.

Secularized motives

For a better understanding of the above, let us briefly describe the motives that have inspired man in his technological endeavours. They indicate the religious nature of the major influences affecting technological development, since man has placed his fundamental trust in technology.

Ever since Francis Bacon said that knowledge itself is power, man has considered scientific knowledge to be the very gateway to universal progress. Of course, this idea gave him a tremendous incentive to increase his knowledge, for science and technology became the means of concentrating on an assured future. Attention shifted from mankind's immediate anxiety, deficiency, distress, and suffering to the achievement of general prosperity for the whole human race. The limitations in this attitude are reflected in Bacon's statement, ''To dominate nature, we must obey her.'' The striking element in this statement is that nature was seen only through the spectacles of natural science. Moreover, obedience in this context implied no more than subjection to the laws of nature. Thus the task of technology was seen as shaping the force and movements of nature to gain ultimate control over it. Thus responsible obedience was reduced to shaping nature technologically by using the laws of nature. The peculiar, intrinsic value of nature itself did not enter the picture, with the result that indispensable ecosystems have been destroyed.

Bacon's second remark also indicates that technology will not be unfolded normatively; instead, its character will be stifled and it will become destructive, not only of the environment but also of life on the social, economic, aesthetic, juridical, and ethical levels. Inherent in Bacon's view is the idea of man thinking entirely and exclusively in technological-scientific terms, an idea that implies absolutized theoretical domination and control of our world and its future. Modern technology must assist in realizing this ambition.

Engineers, in particular, have often been inspired by the idea of technology for its own sake. It is an idea that leads them to contend, among other things, that whatever can be made should be made. Any adverse side-effects produced in the process, such as noise from supersonic aircraft, will need to be taken in stride. Only when side-effects assume unacceptable proportions are technicians prepared to intervene and cure such ailments.

Such a concept of scientific knowledge, however, leads to a one-dimensional, technologically streamlined society. The quest for universal progress is frequently accompanied by an unchecked greed for power, if not openly, at least covertly. Man is eager to break down all barriers in time and space. He is so occupied by this ambition that he becomes a reckless devotee of the development of science and technology. Infatuated with it, he finally surrenders and throws himself and his future on the mercy of technological development.

These various motives imply a common view of nature, namely, that nature is inanimate, insensitive, and open to unchecked interference. It follows that wherever nature proves too stubborn to comply willingly, it is neglected, as for example, in plant and animal species that have become extinct. The quest for progress and the desire for power leave room only for the utterly pragmatic and utilitarian desires of man. The technological *utility* that nature offers is the one thing that deserves attention regardless of the cost to fellow creatures. Sovereign and autonomous man has become conscious of his unique position in creation; at the same time, he has perverted this position because he fails to observe the normative restrictions it entails. Thus he chooses to abuse nature rather than to manage it according to his original mandate. The normative relationship between technology and nature is broken. Man uses technology so that nature is exhausted prematurely, while everything that does not fit into the scheme of technological control is wiped out.

Instead of promoting harmony between technology and nature and thereby unfolding nature according to its meaning, man interferes in nature in such a way that he devastates it. This process will have sweeping results, especially if technology is so caught in the grip of economic powers that profit-making becomes the only valid criterion in the development of technology. Man will then have to discard anything that increases costs; he will exploit nature carelessly, turning it into a refuse heap.

The pursuit of progress and power has produced relatively short-term successes. Nonetheless, there are serious drawbacks, for with neither foresight nor responsibility, man has developed a technology that threatens his survival; he has become the victim of technology instead of its master. Man's inspiration was his vision of total happiness and welfare achieved by his autonomous efforts and aided by cooperation between technology and economic powers. In many ways, he realized just the opposite of such dreams by shaping a world where threats continue to grow in intensity and proportion.

At bottom, the quest for domination, progress, and power has been *secularized*. That is, western man increasingly has attempted to subdue the world even as he denied that God exists or remained oblivious to God's works in this world. Western man acts on his own imagined strength, according to his own insight, and at his own discretion. The only decisive criteria he accepts are material welfare, avarice, and his compulsion to produce and accomplish. He is, as it were, possessed by a nearly demonic passion to dominate and control, urged on by his dreams of the ultimate fruit of his labours: a paradise, a utopia in which he will surely find complete happiness.

Meanwhile, technology has become an idol. The Swiss author Donald Brinkmann says that faith in technological salvation has replaced christian eschatology. Philosopher Oswald Spengler agrees and says that the dominant motive in technology is "the desire for a small, self-created world that reflects the large one because it moves on it own power and obeys the human hand only. The past and present Faustian dream of the inventors is to be God, a dream that has generated the design of all machines." Spengler further states that "technology is eternal and everlasting like God the Father, it redeems mankind like the Son, and it illuminates us like the Holy Spirit."

When men lose their "vertical" orientation, they leave themselves without a proper "horizontal" perspective for this world as well. When men autonomously determine the laws for technological development, they fail to respect the meaning of created reality, i.e., that its meaning cannot be found within created and temporal reality itself, but only in God, its creator and life-giving upholder. Hence they assume that all things exist solely to satisfy the desires of self-centred, technological man and ultimately turn technological strength into a destructive power, as today's environmental problems clearly demonstrate. The ancient legend of King Midas illustrates the point. When Midas was allowed to beg a favour from the gods, he asked that all the things he touched be changed into gold. Like modern economic-technological and materialistic man, King Midas fell victim to greed. Within a few days, he turned to the gods again, asking them to reverse their favour, because literally everything he touched, including his child and his food, had turned into lumps of gold. Even with all that gold at his disposal, Midas was not happy because he could not satisfy his deep personal hunger for love, and he was doomed to starve physically and spiritually.

The moral of this story is that the idols of modern superstition indeed behave as proper idols. They are merciless in their willingness to grant man's every request instead of only what seems to be best for him. Nothing man asks will be denied, even if it costs him something else that he may need. This kind of idolatry ruins life, including the life of cultures. Our technological world amply demonstrates man's basic desires and his deep-seated motivation. The imminent destruction of the environment is a woeful reflection of man's egoism and materialism. Thus today's environmental crisis complements a fundamental spiritual crisis. The Spanish philosopher of culture Ortega y Gasset anticipated this long ago when he said that a man who is forced to live believing in technology and nothing else will lose his meaning. "It is for that reason that these years are the most intensely technological years as well as the most depleted years the history of mankind has ever seen."

Alternative approaches

The implication of what I have been saying so far is that the environmental crisis is basically neither a technological nor an economic problem. It is first of all a *religious* problem. Therefore, a solution must be sought in that direction. But first I will take up two views that cannot offer any help because of their reactionary origins.

The first is the message proclaimed by the true counter-culturalists: that man needs to wake up to his natural origins and adopt a life style accordingly. It is a call for a return to nature, back to the nurture and care of Mother Earth. We are told that we must learn to conduct ourselves as true sons and daughters of nature.

Secondly, some people have also suggested that the pollution of the environment can be halted by drawing attention to the divine character of nature. This is basically a pantheistic idea. Intervention in divine nature (with modern technology, for instance) is a sin that will not be left unpunished and will bring about inevitable and enormous calamities. This view betrays the influence of eastern thought.

Both views are opposed to modern technology as it has developed today. They also express a romanticized view of nature. Man is not considered the master and controller of nature, but an inseparable part of nature (although he has been inserted into it). If these views should ever be worked out in society, the consequences for mankind would at least match those of an unchecked technological development.

For without the possibilities offered by modern technology, life would become impossible for many. The weak could become extremely vulnerable since they would be unequipped to deal with a hostile and unyielding nature. No meaningful solutions to our problems can be found by taking a position that simultaneously favours nature and opposes culture. Such an approach lacks sensitivity to nature's two faces — the benign and the hostile — and to the unique position of man; nor does it recognize the meaning of history.

In the face of our idolatry and misjudgement of technology, it is important to recall man's original mandate concerning technology. However we cannot fully appreciate this mandate unless we first recognize and confess the true Authority, who gives life to man and who crowns him with honour and dominion. Thus man is accountable to God for responsible stewardship that stresses service instead of autonomous control. His technological endeavours are subject to norms that are supra-human and, therefore, not subject to whim or caprice. God calls upon man to cooperate in the progress of history and in the disclosure and development of creation.

The Gospel — the good news — offers a timely perspective here. Christians especially need to begin with self-criticism, for too often they have simply adapted themselves to the dominant materialistic tendency within western culture, at times even defending their weakness with texts from the Bible, thus compromising themselves in a frightful way. We know from the Bible that through the work of Jesus Christ, God has reconciled himself with his creation, unsettled as it was by the effects of human sin. This reconciliation implies that man must resist a self-centred and avaricious exploitation of nature. The abuse of nature cannot be in harmony with responsible stewardship in God's creation. Any human claim to the ownership of nature must be denied, for the materials that constitute life can never be anyone's property. This is amply demonstrated by the functioning of ecosystems. ''Building materials'' have been used by a chain of succeeding generations of living beings throughout the centuries. This should also be possible in the future, unless this generation of mankind is to disrupt this chain by idolizing a technology based on an absolutized scientific method.

To avoid disrupting the chain, man needs to recognize that scientific knowledge will always be of limited scope, since it is abstracted from the fullness of reality. Reality cannot be fully understood through scientific knowledge alone. By refusing to admit the limitations to scientific knowledge, we have dislocated the complex coherence within reality. Multi-disciplinary cooperation may bring about a partial improvement in our efforts — partial, because we will not gain a comprehensive knowledge of the fullness of reality simply by adding the abstract and limited knowledge of various aspects. We will be able to restrain the devastating powers unleashed by modern technology and to restore its constructive and benevolent properties only after we acknowledge both man's individual and communal responsibility to his creator and, therefore, for his fellowman, for his fellow-creatures, and for nature. Accepting this responsibility clears the way for recognizing the limits of the human capacity to command our complicated, varied, and dynamic reality. Such responsibility should also lead us gratefully to employ the various kinds of abstract knowledge. Science cannot be expected to integrate these forms of knowledge, since such integration is part of the responsibility of man himself.

This is by no means an easy task. Anyone aware of the difficulties it raises will proceed cautiously and with restraint, especially where we are concerned with altering reality, as in technology. But this insight will lead us to handle nature prudently and carefully. It will also help us maintain the delicate balance between the natural environment and living organism. On that basis every suggestion that leads to solving the problem of environmental pollution should be regarded positively. The discussions of a production process that involves recycling need to be changed into action as soon as possible. Moreover, we need to take seriously the calls for austerity, for new forms of asceticism, for decentralization of industry, for an end to unbridled consumption and the technological compulsion to produce and accomplish. We also need to pay attention to those who suggest that we change societal structures, work for more efficiency in political decision-making, break up elite groups, expand the effort to democratize decision-making processes, reinforce community consciousness, and place restrictions and normative guidance on our dynamic technological development. All of these are positive suggestions that deserve our full support.

If, however, the inspiration for these proposals is again the subjective desires of autonomous man, they will have only limited and short-term effects. For subjective religious attitudes as a source of inspiration will always be one-dimensional.

Let us grant for a moment that the pollution of the environment can be almost entirely eliminated technologically, assuming the cooperation of economic forces. If the basic changes are limited to using technology merely to deal with symptoms, then technology, bureaucracy, and existing organizations will probably continue to grow in power at the expense of individual freedom and communal responsibility. Man will then become the prisoner of his own technologically streamlined society, which will become a universal prison-house, in which historically, socially, and ethically meaningful life is constricted and finally choked. Life will then hold little more purpose than it does for a fly caught in a bottle; mankind will be doomed to die. Our quest for solutions will only produce more problems as long as we pursue them within a strictly horizontal framework and take no account of man's essential vertical orientation to God, who has indeed issued clear and normative guidelines for a development of technology that harmonizes with nature.

That means, concretely, that technological development should also unfold in a social sense and that everyone involved in such development bears a measure of social responsibility. Further, we should not close off the economic side of technological development, but we must stem any development that turns profit-making or the compulsion to consume into absolutes. We need to prevent squandering of resources and goods and to promote a frugal management of nature, even if such a move should be accompanied by a lowered economic growth and by restricted consumption.

In addition, if we halt the accumulation of garbage and industrial wastes, we will be demonstrating our sensitivity to the aesthetic disclosure of technology. Technology should not be permitted to spoil nature but must be developed in harmony with it. A clean and protected environment is also a judicial concern. Wherever nature is abused, the courts should use their power to intervene and dole out suitable punishments to guard against further pollution and mutilation of nature. In so doing they will ward off the threat that a ruined nature poses.

Nature is dislocated wherever man's ethos assumes an erroneous content because those involved in the development of technology arrive at their point of reference by making something within created reality an absolute. In that situation, they do not consider their actions as part of their service to God who rules over creation in Jesus Christ. They exchange the perspective of the kingdom of God for shortsighted and relatively short-term success because, being unreceptive to the Transcendent One, they also lack vision. It is precisely such receptivity and vision that make a meaningful perspective for our culture possible.

The most fundamental basis for a different attitude towards

nature must come from the recognition that we are living in God's creation, a coherent whole in which every part has its peculiar value and place. With that recognition, the ''vertical'' relationship between God and man will be restored. Only then will we be able to establish the kind of perspective that will allow us meaningfully to cultivate the ''horizontal'' relationship. We will be able to participate communally in the normative meaning of everything, including the meaning of science and technology, rather than communally to pervert life until man himself is driven to curse. We will be able once again to enjoy nature and to see that all creatures great and small are the letters that together form one great book telling us about the power of God the Creator, of Jesus Christ the Redeemer, and of the Holy Spirit the Restorer.

40 The environmental problem

Reflections on the technological-scientific culture

It is well-known that the study programmes designed for prospective engineers leave little or no room for reflecting on technology and its possible consequences for culture. The emphasis in technological education is on the continuing development of modern technology; only marginal room is left for reflecting on that development. Even then, it frequently amounts to no more than an option for engineering students.

As a result, engineers emerge from their training naively engrossed with the idea of permanent progress brought about by technology. Thus they think that their efforts will contribute to a future in which technology will eliminate many maladies and in which material welfare will continue to grow. Whenever engineers are confronted with problems in shaping such a future, they usually lack the time to reflect on their causes and automatically summon the help of science and technology to solve them. In his daily work the engineer is often so preoccupied with the impressive results of his work that he has little interest in dealing with philosophical questions concerning technology, even if he had the time.

Furthermore, the engineer's role in society is increasingly limited to promoting the development of technology from his position on a team of colleagues. Being a specialist himself, he grows more and more dependent on other specialists because the making of a technological design has become an immensely complex project. This trend is reinforced by advances in the methods and tools of technology, especially the computer. As a result, the individual engineer does not have the insight or ability to help guide the development of technology along a particular course. Given the meaning of modern technology for culture, the responsibility of the engineer has greatly increased; yet his training and his work situation make it difficult to him to carry it out. Unfortunately, the engineer often does not see this himself.

During the last few years, however, there have been a number of substantial changes in society and in engineering circles. For a long time, technology was thought to be the only means of moving towards a good future. But today people who live in the midst of the technological-scientific culture are troubled by persistent doubts. Many engineers refuse to ignore the problems of this technological age any longer. Among them there is a keenly sensed and growing pessimism concerning not only technological advancement but also the heart of technological development itself.

An atmosphere of crisis prevails in engineering circles. Although engineers realize that technological development has meant tremendous material enrichment to many, especially in the west, they also recognize that technology has developed into a power dominating the human situation. Technology has so thoroughly unsettled our society that we can no longer evade an important question: is this development of technology moving in a right direction? Engineers are now asking questions which were not posed until recently. Reflection on those questions should be part of the prescribed programme of the engineering student.

I would like to contribute my share, taking reformational philosophical thought as my point of departure and focusing especially on the influence of science on technological development. Although this influence is extensive, it is not often examined critically. Instead, people assume too often that incessant specialization and a high intellectual level are the prerequisites for coping with impending problems. I would like to challenge this view and to examine critically the prevailing view of the relationship between science and technology, its deepest foundation, and the motives at work behind it.

The problems of the technological-scientific culture

The central significance of science is demonstrated from the outset by the various problems surrounding the technological-scientific culture. By way of explanation, let me briefly summarize the most serious problems of that culture.

The first problem area concerns the position of man in modern technology compared with his position in technology of earlier days. In the technology of the trades, man's productive activity was determined by his eyes, his hands, his feeling for materials, his fantasy, and his ability to shape. His physical and intellectual capacities were involved directly. Quality, particularity, and uniqueness characterized his work and the results of that work. This situation is entirely different in modern technology, where quantity, rate increase, and volume characterize man's work. Work has become quantified labour which must continually grow more intense.

At the same time, working man has become part of a process which is objective, impersonal, and insensitive, a process in which cold business gains the upper hand. The result is hardened and indifferent workers. Their attitude is quite understandable, since man feels himself locked within an overwhelming, gigantic mechanical process that drains his resources. Even man himself is set aside when complete automation is introduced, and he remains at the sidelines as an unemployed spectator.

An existential fear often takes hold of him in this situation, a fear that understanding cannot relieve. In fact, man grows more afraid the moment he realizes what is really going on: the labour process, in which he is but a small cogwheel, requires unchecked and gluttonous drains on the sources of energy and on the reserves of natural resources. The drain goes on as if they could never be depleted, whereas these sources and reserves are, in fact, finite. Therefore as technology grows to even greater proportions, it becomes clear that it must also come to an end; and the more the dynamics of the process are stepped up, the sooner that end will come. This will inevitably lead to enormous disasters in a culture that has become highly dependent on technological development. Even at an early stage in technological development, Romano Guardini saw that unlimited technological development harbours chaos as the final outcome. The dictates of technological perfection disguise an enormous abyss. A catastrophe of infinite proportion and swiftness threatens to put an end to finite physical reality.

Moreover new frontiers of scientific technological advancement are already inspiring fear. The development of nuclear energy, for instance, has been sold to the public with promises of a guaranteed energy supply to meet increased demand. Of course, we will also have to endure the many unsolved technological and political risks, whose magnitude is difficult to estimate. The question of whether or not man can effectively control the elementary forces released by matter continues to plague us.

The case of computer technology is quite similar. A sober analysis indicates that the computer works fast and accurately and that its results will never go beyond the programmed instructions. Yet people's fear of growing more dependent on the computer remains real because the computer operates independently of man himself, because its results contain a limited element of surprise, and because the user is not necessarily the programmer. Moreover since the user changes again and again, he cannot know by what set of criteria the computer works; he is forced to surrender himself in trust to the dictates of the computer. This problem will be aggravated when the self-adapting and self-reproductive machines, predicted by computer specialists, are introduced in the future. A computerocracy is imminent when these machines are expected to find their place not only in production processes but also in economics, politics, and government. Manipulation of data, especially as derived from data banks, and manipulation of people could well assume dangerous proportions.

43

The threats attending modern technology are particularly striking when we observe the latest military techniques. The most deplorable possibility is nuclear suicide. Many dangers are also associated with bio-technology or eugenics because it is based on biochemistry and biophysics and therefore possesses the built-in possibility of manipulating genetic materials.

Another important problem area in our technological-scientific culture concerns the relationship between man and nature. The small-scale technology of earlier days was integrated with nature, but today a breach exists between man and nature. Aided by modern technology, man continually interferes with nature on a large scale, until the given coherence of nature is disjointed or even broken up entirely. In addition, man interferes so frequently that nature is not given a chance to repair the damage. Add to this the combined adverse side-effects of modern industrial technology, of traffic, and of industrial agriculture and we may conclude that the danger of environmental collapse is even more real. We draw from nature more than it is able to produce, and we discard more waste products than it is able to break down, not to mention those that cannot be broken down. The possibility of completely destroying the environment is becoming very real indeed.

In summary, the threat modern technology poses can be looked at from two angles. From one angle we see the gigantic, massive structure of technology and the frontiers it advances, while from the other we see the technological disruption of the coherence in nature and the alienation of man and nature.

All told, it is understandable that many people look on the explosive and ambitious development of technology as an autonomous development, particularly when they take into account its socioeconomic context. The research needed for the development of technology occurs in mammoth organizations and within the so-called military-industrial complex, often in secret. Should the state ever take over this task, we would experience technological development as an inescapable and unattractive fate. Our specialized and massive technology no longer meets actual needs and desires, which have given way to goals systematically fixed by a system-technology founded on growth-ideals. The unscrupulous use of technocracy and the dehumanization it brings continue to grow, both in proportion and in complexity.

When we look at the influence science exerts on technology, we can better understand the problem that has arisen in man's relationship to technology and to nature and its socioeconomic coherence. Because modern technology and the technological-scientific method of designing rest on a scientific basis, the characteristics of natural science project themselves into the production process and the results of technological activities. These characteristics emerge wherever technological products are introduced, leaving a decisive mark on various cultural sectors. As some cultural sectors dominate others, the characteristics of science also become cultural characteristics. [1]

The characteristics of natural scientific knowledge are these: it is *universally valid, abstract,* and *remote.* That is, it is a reduced and limited knowledge of reality because it is a knowledge of certain aspects of reality only, abstracted from the fullness of reality. Further it is *enduring* and *perpetual,* and it shows a *logical coherence.* These features of scientific knowledge are frequently mistaken for full and complete knowledge of reality. However integrated reality is marked by the uniqueness, the coherence, and the interchangeability of everything. Therefore a tension arises between scientific knowledge and reality as a whole. [2]

The features of scientific knowledge — universality, abstractness, and perpetuity — will also become technological features because science and technology are so closely associated, even joined. Since scientific knowledge is based on rational, logical abstraction and exhibits a purely logical coherence, the products derived from it will also feature this logical coherence. As a result, the absolutization of scientific, logical knowledge by technology is attended by an unprecedented determinism. The mathematical and mathematical-physical laws for reality are then held to be the true and full picture of reality.

Developing a technology based on such absolutization will inevitably lead to enormous control over people. There will be little room for man to use technology creatively to shape his historical situation. Technological activity becomes regulated and constricted; hence man is robbed of his inventiveness, of his responsibility, and even of his freedom. The upshot of this is the rise of a monotonously standardized technological mass-culture, in which people are equalized and levelled out to be average. That is to say, their work situation and their consumption patterns can be interchanged, statistically managed, and manipulated.

Modern technology has become part of all world cultures, moulding these cultures into its own uniform and monotonous mass pattern. The characteristics of technology are the same everywhere. Building construction, for example, is identical everywhere, whether this be in the Netherlands, the United States, Japan, or a developing country. Meanwhile, this process has dislocated and even eradicated many ancient cultures.

Moreover, the technological-scientific culture itself is in danger of being internally torn and segmented. However surprising this may seem, it is the inevitable result of scientists in each specialized area projecting entirely different views onto reality. The problem is the result of indiscriminately applying scientific knowledge.

Consequently, no genuine cultural integration takes place, unless we mean technological-scientific integration, which implies a tremendous reduction of cultural life.

We can already see the impact of such reduction as work grows monotonous, as nearly everything becomes standardized, as human life is broken into distinct and even isolated units, such as home, work, and recreation, and, finally, as we disrupt the coherence in nature and break the intimate bond between man and nature. We can be sure that in the future technology will expand still further, for it displays a tremendous dynamic. But this expanding grip on life will stifle and restrict life. Ultimately, it will eliminate the rich variety, complexity, and stability inherent in human life.

Already our culture has become a very unstable mono-culture with tyrannical tendencies because of the bias and the dynamics of the technological-scientific culture. An obvious indicator is the emergency created by a sudden ''energy crisis'' and the socio-economic difficulties it caused. This danger is still with us. But the totalitarian character of the technological dynamic makes it very difficult to alter this established pattern, in part, because people do not agree on how the culture should be changed and also because the force propelling today's culture is so strong. While this force has brought material relief and progress for humankind, we do not seem able to harness it without risking immense catastrophes.

I can only conclude that the problems of the technological-scientific culture can no longer be described as incidental or acute; rather, they are structural and chronic. They are an expression of an emergency situation that is growing.

Initially, man seemed to use technology to subdue natural forces and to deal with emergencies — however resistant and unpredictable; still, he found this task difficult. With natural forces subdued, man is now faced with the problems accompanying cultural forces. The road to mastery over natural forces, for all its hopeful beginning, seems to end in the morass of a cultural crisis, one that is serious enough to threaten both man and nature.

A look at some philosophical views of technology[3]

The basic problem in the technological-scientific culture rests in the view that correlates science and technology. However, this problem is not often stressed in today's philosophical assessments of culture. Surprisingly, it is the pragmatists and the positivists — the philosophers who communicate readily with engineers — who typically ignore this problem. They take their bearings from the existing coalition between science and technology, evaluating its development positively and interpreting it as a confirmation of human strength and ability, a great stride forward on the road to general welfare and prosperity. Whenever obstructions occur on this road, these philosophers frequently issue an uncritical appeal to science for a solution. When the problems assume the proportions we know today, they do not hesitate to advocate a total domination of culture, aided by modern systems theory and cybernetics.

These philosophers are confident about the future, assuming that they will enjoy the continued support of the current development of science and technology. Infatuated with the dream of complete mastery of the future, they resist religion (Christianity in particular) and those philosophers who demonstrate their dependence on and guidance from the *Hinterwelt* (meta-world, transcendent reality); they must resist them because they would prevent them from realizing their ideas. The closed world view of the technological-scientific culture is a *conditio sine qua non* for pragmatists and positivists.

The orthodox Marxist also accepts the importance of the development of modern, scientific technology, stressing man's technological achievements and contending that man can reach his ultimate goal of total freedom by developing the potentialities of technology. Unlike positivists and pragmatists, the Marxist is aware of the many problems and dangers involved in this development, including the possibility of alienation and loss of freedom. But he remains confident and argues that continued socioeconomic development based on technological expansion, together with the inevitable revolution, will ultimately eradicate the growing shadows of alienation and restricted freedom. They will open up the kingdom of total freedom at the dawn of a new day, a kingdom in which humankind will reign collectively as lord and master over the work of his own hands.

Positivists, pragmatists, and orthodox Marxists all have great respect for science as a means to control. However their assessment of its effect varies because of underlying differences in their views of society.

The Marxists do not begin with either man's freedom or the production by free enterprise within which technology is to develop. Instead, they emphasize that technology can benefit the revolution and the liberation of humankind only if it is guided centrally and if goods produced are judged by their value for consumption instead of exchange. In other words, they propose a centralist technocracy.

In passing, it should be noted that, in view of the growing gravity of the problems, positivist and pragmatist philosophers have had to reassess their views of increased state intervention. Thus many have correctly concluded that marxist society and non-marxist society are beginning to resemble each other.

The existentialists and the personalists, who are keenly aware of the problems facing the technological-scientific culture, take a transcendentalist approach in their analysis. They emphasize the transcendental influence on all experience, interpreting the ongoing development of the technological culture as a menace to the human subject, especially to his personal uniqueness, to his freedom, and to his individuality. They oppose science and technology as forces that are autonomous and anonymous. To be sure, they oppose not the big problems of the technocratic society but science itself, arguing that the scientific method leads to oppression and the suppression of human freedoms.

The existentialists and the personalists make quite an impressive protest, but, in view of their presupposition that science is an autonomous original force, we have to ask whether they are capable of pointing out a meaningful perspective for the future. They submit themselves to the present development by sheer necessity, look to the past nostalgically, or attempt to escape from the culture in a transcending retreat towards freedom; thus they rise above the problems looming over science and technology. But even this freedom is threatened continually. While they reject the urge to dominate, which is so common to technological-scientific thought, they rely on their own thought to help them rise above that urge and to enter a space of spiritualized freedom. However, they must often repeat their spiritualized flight as their concern for external aspects continually grows.

In their evaluation of man's position in society, the neo-marxist revolutionaries agree with the transcendentalists. Although they do not consider science and technology to be autonomous powers, they fail to submit science and its relationship to technology to a basic and critical examination. Their criticism amounts to attacking economists and politicians, charging them with elitist tendencies and a willingness to enlist the service of science and technology to their own advantage. Inspired by the vision of a future utopia in which everyone will be free and happy, they want to transform present society until it conforms to their model. If their revolutionary ideas are realized, they will have sweeping consequences for the development of science and technology, including their place in economics and politics. For the neo-Marxists would give priority not to the solution of practical *problems*, but to the realization of practical *purposes*, a process that may not involve restraint (v., Habermas).

The neo-Marxists offer revolutionary resistance to the ideology of technocracy so as to give human authenticity a chance in everyday practice. In their view, man is not primarily the creature who works and produces, but the *homo ludens* who plays, relaxes, and has a good time. Thus they hope to fulfil human existence by meeting man's need to enjoy life and to express his passions fully, goals that were previously nearly impossible.

The countercultural critique

No matter how important and penetrating the critique of the transcendentalists and the neo-Marxists may be for problems of the technological culture, in this study I would like to devote special attention to those people who have spoken for the so-called counterculture, particularly Theodore Roszak. The ideas advanced by these people and their followers have served as a marginal undercurrent within western culture since the days of Romanticism. Recently these ideas broke through to the surface and occupied the minds of many, and not just the young.

The representatives of the counterculture formed a front with the transcendentalists and the neo-Marxists in their serious objections to the position of science and technology in modern culture. With the transcendentalists they pleaded for a return to authentic thought and for its renewal. However, when they advocated ways of applying this renewed reflection in the culture, their proposals did not correspond with those of the revolutionaries.

For one thing, the revolutionaries and the transcendentalists differed in their concept of revolution. The neo-Marxists proposed a revolution of *society*, a transformation of societal structures, and therefore a change in the function of science and technology. The spokesmen for the counterculture, however, advocated a *spiritual* revolution, an internal revolution of the mind. This amounted to an attack on science itself, for they critically examined the structure of science and the peculiar course science and technology have taken in western culture. Roszak stated that the rise of the counterculture must be seen in the light of the meaninglessness and utter despair produced by the technological culture and the mad pursuit of the excessively specialized sciences, which tended to dissolve the integrated character of human knowledge. Countercultural critics called for communal awareness, authenticity, and meaning in the midst of prevailing social upheaval, alienation, and meaninglessness.

If the human mind is to be cured and healed, critics said, it first needs to be aware of the cause of its illness. Roszak thought that this illness originated with the rise of modern science, particularly stimulated by the thought of René Descartes and Francis Bacon. From its start in the seventeenth century, science has developed into the final and highest court of appeal. The technological-scientific culture is based on such science, used as an instrument to gain control. Therefore this culture now looms large and looks like a totalitarian, artificial, and abstract culture having a linear, horizontal dynamic. Roszak emphasized that the groundwork for this development had been worked out by Judaism and Christianity, since they used a linear view of history, which was advanced especially by later Calvinism. With the rise and development of science, the knowledge of the *natural* sciences has been accorded a superior status. All other forms of knowledge are considered inferior to the lucidity of objective scientific knowledge.

As I pointed out earlier, scientific knowledge is limited in its scope; it is the result of the human mind observing strict and objective limits and leads to a partial, reduced view of reality. It is precisely this constriction of the mind that provides scientific knowledge with its apparent strength, for within this uncomplex, uniform, and single view, all of reality is reduced to the categories of mathematical and physical laws. Thus some can pretend that such reduction clears the way to a true and adequate picture of reality that can be manipulated at will. Roszak called this the *religion of science*. This religious veneration of science helps explain the scientist's willingness to sacrifice his energy and his triumphant claims that he has grasped reality through the eye of reduced knowledge. Thus reductionism is on the increase, and more is lost than gained in the name of progress. The end is a shrunken technological-scientific view of the world, that can only lead to a meaningless and nihilistic culture.

The countercultural message was addressed to this culture and advocated a salutary return to the mystical and visionary sources of authentic culture. The apologists of the counterculture wanted to keep the objective, constricted, and alienated mind at a distance and return to a dedicated, visionary consciousness based on feeling rather than on reason. This, they suggested, is the first step towards liberation from the total alienation caused by the technological-scientific culture.

They hailed other forms of knowledge, such as those provided by imagination, intuition, wisdom, mystery, inspiration, ecstacy, contemplation, meditation, myth, gnosis, passion, the unspeakable, the mysterious, and the holy. Their aim was not to know as much as possible, but to know as profoundly as possible, to move away from continual abstraction towards more deeply meaningful, "transcendental" knowledge.

They suggested that people appreciate the nonintellectual faculties of the human personality, that are nourished by visionary sparks and by wholesome human experience. Only by returning from expertise to wisdom, they said, could men and women participate in the good, the true, and the beautiful. These newly enlightened thinkers called for a mind that participates in the divine; they sought the natural mind, perfect in itself, as opposed to veneration of science and technology. They preferred child-like receptivity and simplicity to the mania for self-esteem common to men of intellect and culture. They pled for union of heart and feeling as opposed to the gap between heart and intellect that troubled technological-scientific man. Modern man, they said, not only controls his technological culture but is also controlled by it at the expense of his multidimensionality, of his wisdom, and of his independence.

The values and norms of the counterculture have formed a striking contrast to those of western culture since the scientific revolution of the seventeenth century. In a sense, countercultural apologists advocated a return to a far earlier time. In effect, they replaced a linear view of history with a circular or cyclical view, never mentioning cultural continuity and progress. Of course, they did not reject science and technology altogether.

Science and technology had their place insofar as they were required for the survival of mankind, but the countercultural apologists insisted they must be cut back to size.

The distinguishing feature of the counterculture, therefore, is that it offered a limited, human scale, that was varied even in its societal forms, rather than monotonously identical at every level. It stressed the organic above the mechanical, simplicity and frugality above abundance, and meaningful and delightful labour above production-oriented labour.

Thus countercultural advocates called attention to many important issues often neglected in the past. Much of their analysis of the technological-scientific culture was correct, especially their critical observations concerning the structure of science and their emphasis on the religious backgrounds of the issues they raised.

However when the apologists of the counterculture reacted against the adulation of science and technology, they reduced them to a mere requirement for the survival of mankind. Since they did not recognize a cultural mandate, they have been unable to shape an alternative direction for present culture or even to deflect its current direction. In effect, the ideas they introduced into public discussions could only sponge on a culture that is internally torn and segmented.

The philosophy of counterculture took an extreme position on the continuum between rationalistic technological-scientific culture and the irrationalistic reactions it evokes. Since the counterculture offered no alternative for cultural development, it could offer little or no actual resistance against increasing cultural dislocation. It may well have had a completely opposite effect because no real resistance could be provided. If the spirit of the counterculture were to dominate the heart of the technological-scientific culture, the latter would be undermined from within, for it would lack the courage to accept cultural freedom and responsibility. Our culture would then be confronted with imminent collapse, which would have far more serious consequences than the hazards it has been shown to harbour today.

The motives behind technological development [4]

Present cultural development must be slowed down, modified, and redirected. But how is this to be done? To answer this question, we need a clear view of the motives underlying and driving this development. We also need to understand the fundamental religious background of these motives and of the cultural development they spawn. I agree with Roszak that the major motive in western culture today is man's will to master and control, combined with the idea of technology as applied science. After all, the dictatorial character of science can be seen most clearly in its application, that is, in modern technology.

As man attributes absolute power to science and to the technological-scientific method, technological development appears to reflect scientific knowledge. This leads to an extremely scientistic technology, particularly on the design level. As a result, human creativity recedes or even disappears altogether. Thus all invention is stifled, and the possibilities for redirecting our technological efforts are reduced. As I said earlier, scientific knowledge is enduring and perpetual because it is knowledge of a fixed and limited subject matter. In turn, this continuity is projected into technology, which becomes fixed and limited also.

Thus domination by rationalism has made the development of technology not only dynamic and immense but constricted as well. Innovative work receives only a very small place because people influenced by rationalism singled out the prevailing technological development as the only proper one. Thus the development of technology becomes stifled and rigid in its rational determinism and relentless logic. However, our obsession with technological accomplishments blinds most people to the narrowness of our current technological development.

However the human will to control by technology is not the only motive operating in western culture. In fact, it is intertwined with various other motives, some of which are corrective, but most of which buttress the dominant trend. There is, for instance, the motive of *technology for the sake of technology*. Let us call it the imperative of technological perfection, demonstrated most notably by the exclusiveness of the ideas proposed by engineers. Whatever can be made and perfected must be made and perfected. This leads to unchecked and meaningless technological power, which engineers claim they master and control but which in fact victimizes them. Although the motive of technology for its own sake often anticipates progress, it actually does just the opposite: technology dominates man with absolute power. Even nature and culture cannot escape the menace posed by this technological power.

Frequently, rationalism is also allied with the motive of technology as the servant of economic powers. These powers dominate the development of technology, often turning profit-making into an absolute good. Admittedly, this makes it possible to interrupt technology's current development, but it also means that technology is developed without the norms which ought to be applied in the first place, such as responsible management of the environment. Under the influence of this economism, technology ceases to answer to its essential meaning. It no longer serves actual needs but creates artificial needs and superfluous products. The result is an authoritarian technological development that leaves behind a trail of waste, pollution, and destruction.

These motives are especially influential on those who are actively engaged in the development and social direction of technology. Those active elsewhere may be influenced by different motives: for example, the belief that technology is a neutral and autonomous power. Yet they, too, frequently close their eyes to the dangers of technological development. Blinded by an insatiable desire for welfare, they willingly adapt to the prevailing development in order to derive private benefit from it.

Thus our search has taken us to the fundamental basis for these various motives. Could it be that the large-scale problems and threats of a large-scale technology arose because these motives are based on large-scale human pretensions?

The spiritual or religious background of the technological-scientific culture

Roszak was correct when he went back to thinkers like Descartes and Bacon for his evaluation of the technological culture. For it was Descartes who pleaded for the autonomous and self-sufficient position of man: man as subject was the measure and centre of all that is. And Francis Bacon was a stimulus particularly to the practical consequences of this view, for he taught that man is capable of realizing his autonomous position with the aid of technology and science. Technology can be used to control nature and to create the kind of culture that surpasses all restrictions of time and space and that obeys the hand of man In the modern age, man is caught in the clutches of the infinite, committing himself to the limitless possibilities of science and technology, particularly after the age of Enlightenment.

Earlier I stated that positivists and Marxists consider Christianity an obstacle to the growth and progress of the technological-scientific culture. Roszak, by contrast, says that this culture originated in the judaeo-christian tradition and was particularly advanced by later Calvinism. Roszak attributes the present cultural problems to Christianity because the history of Christianity produced thinkers like Descartes and Bacon and also the pretensions of the Enlightenment.

Such contradictory views indicate a problem that deserves our attention. In my opinion, both views are wrong. Undeniably, the judaeo-christian and especially the reformational view of man's unique position regarding nature was a decisive factor in the development of the exact natural sciences. But more has to be said about the manner, condition, extent, or purpose of this unique position of man. For example, did the authority man received to exercise dominion in created reality imply unrestricted sovereignty? People draw this conclusion too readily when the biblical mandate found in the beginning of the book of Genesis is taken out of context. It is a mandate whose hallmark is service. The Bible clearly points out that man must resist the human temptation to misuse his given mandate to manage and to have dominion in creation.[5]

It is the philosophy rooted in the thought of Descartes that has focused its attention from the outset on the exact natural sciences. The idea of human autonomy posited in Descartes's philosophy stimulated the development of the natural sciences; this development, in turn, appeared to confirm the idea of autonomy. Since the Enlightenment, this idea has also been made effective by the application of scientific knowledge and by the projection of scientific characteristics into culture. Guardini has shown that when basic religious meaning is replaced by pretended autonomy, the resulting void will be filled by violent, practical, manipulative science. The biblical mandate was continually perverted during the course of the historical development of western society and finally amounted to technocratic exploitation. A passion for perfection and completion has caused western culture to secularize the biblical mandate: man takes his destiny into his own hands, while christian eschatology is reduced to a dogma of horizontal progress in history.

When Christians, at least those Christians who hold a dualistic view of life, associate themselves with this dominant persuasion, Roszak is correct in accusing them of complicity in the realization of our science-infused culture. However, positivists and Marxists are often correct as well in charging Christians with opposing science and technology. They reach this conclusion because they frequently observe Christians who oppose the development of science and technology as a result of their failure to integrate a responsible attitude towards it with their dependence on a transcendent reality. Such Christians neglect the fact that the christian faith embraces values pertaining to created reality.

Indeed, faith thus understood rejects the idea of human autonomy, but not responsibility for science and technology. With Roszak we reject man's excessive pretensions for science and technology, but we cannot therefore support the proposal of the counterculture, since they do not give up the idea of autonomy but merely gives it a different content. Roszak does not see man as autonomous regarding nature. Rather, man and nature belong together in an indissoluble bond with its own point of reference. Hence Roszak's pleas for transcendence should be understood as a restoration of that bond. Despite appearances, the transcendence Roszak advocates is immanent; it remains within this world and its multifaceted reality of man and nature. According to the spokesmen of the counterculture, the limitless mystique of transcendental experience establishes a harmony between man and nature and a harmony of man with himself. In that experience, escape from an alienated and threatened existence in the technological-scientific culture is thought possible.

An alternative
Roszak rejects the prevailing trend of our culture. So do I. But unlike him, I would like to defend an alternative approach in which a transcendental orientation includes responsibility for technological-scientific development.

For a better understanding of this approach, we must take a brief look at the spiritual history of the western world. The most fundamental basis inspiring man's scientific and technological activities has been the idea of autonomy. Within the tradition of western philosophical thought, philosophy has been shaped and developed to support and confirm the pretended autonomy of human thought. Philosophy was thus accorded the function of a religion or a pseudo-revelation. Subsequently, philosophy, in the line of rationalism, which generally orientates itself to the exact natural sciences and modern technology, was reduced in scope until it was confined within the natural sciences and their methodology. In this context, we can readily grasp C.F. von Wiezsackers's remark that scientism as a faith in science has increasingly assumed the role of religion in western culture.

Thus philosophy is used as a stepping-stone to combine the idea of autonomy with science and its methodology. People then begin to find assurance and confidence in a tacit, religious devotion to the scientific method, used in gaining mastery of practical affairs, particularly the affairs of modern technology. Philosophy is even equated with the most characteristic feature of science, namely, abstraction. Accordingly, the scope of philosophy is increasingly reduced and constricted in its continuous adjustment to the nature of scientific knowledge.

As man humbly submits himself to the dictates of this abstract scientific thought, he will very likely absolutize it. He forgets the reducing character of science and suggests instead that science produces complete and concrete knowledge of the whole of reality. Whenever people hold this view of abstraction and wherever in triumphant anticipation of progress, they apply scientific knowledge to project this abstraction into reality itself, the result will be a constricted reality, that is, a reality of what is technologically possible. In fact, it leads to a fragmented distortion of reality and a dislocation of culture.

It becomes clear, therefore, that an uncritical appeal to science as a panacea for the problems facing our culture today will indeed bring temporary relief only. In the long run, in view of the inherently limited structure of science, this reliance on science can only confront us with new and even more distressing problems. In fact, the solution it offers provides us with a more involved and more obscure complex of old and new problems.

View of science

Rationalistic philosophers never concerned themselves with the problem that there is a still more fundamental way of knowing that precedes a conceptually qualified, scientific knowledge. It apparently did not occur to them that there might be a mutual relationship between these ways of knowing. Historically and structurally, the kind of knowledge used by the special sciences is restricted to definite boundaries and is based on a knowledge that is fundamental, uncompartmentalized, and concrete. This pre-scientific knowledge, which consists of both actual and factual knowledge of changes effected by human action, is in its turn guided and kept on course by an original and irreducible ultimate trust. This trust-knowledge constitutes man's point of reference for all his philosophical and scientific endeavours, regardless of whether he realizes it. But in autonomous theoretical reflection, inspired since the beginning of the modern age by a religious faith in philosophical and scientific thought, this pre-scientific trust-knowledge has been thoroughly perverted and falsified. Meanwhile scientific knowledge and its practical application have been accorded superior status.

Reformational philosophy, however, rejects the autonomy of man and therefore the autonomy of philosophical and special scientific thought as well. All reformational thought begins with the acknowledgement that nature, man, and culture are not anything in themselves. On the contrary, it recognizes that man, who, after all, did not create himself, needs a revelation in order to find out who he is, for what purpose he exists, and what is the meaning of the history that encompasses him and all things. Guided by divine revelation, man is in a position to become aware of the origin of everything, of the cause of his distress, and of the two-fold meaning of nature, of human life, of work, of culture, and of history. At the same time revelation enables man to learn about the way of redemption from his distress.

This leads to a knowledge whose content is open to understanding by faith; it is the most profound knowledge, which our thinking cannot fathom. It is a central, deep, and comprehensive knowledge that concerns the very heart of man and that, for man's own sake, should have decisive influence on scientific thought and on technological endeavours. It is a knowledge of acknowledgement, of confession, springing from the heart itself. This knowledge is dynamic rather than static; it is a gift as well as a mandate. It is unfolded continually as long as there is an obedient and receptive understanding of divine revelation. That is to say, when man renounces every pretension to autonomy, he finds opening up before him the very way to life, wisdom, and insight. The content of this insight is made explicit in faith-knowledge. It is neither philosophy nor science, for these are both characterized by abstraction. Faith-knowledge is radical and integral, for it concerns the central, fundamental choice of human life in the very heart of human existence. It shows man which direction he is to take in history, and it motivates him to follow it through.

What does all this mean for science? This question is critical in view of the theme of this essay. It certainly implies a view of science critical of the views widely held. It is also significant for the development of technology.

First of all, the nature of scientific theories needs to be recognized. The characteristic feature of these theories is not that they are objective or value-free, as is so often claimed even today. Rather, these theories have a basic religious foundation, although personal and social conditions may also play a role. For that reason, scientific theories should *not* function as if they were absolutely independent, nor should they be permitted to straight-jacket our knowledge and our action.

A theory starts out from a hypothesis. A hypothesis, which is an expression of human creativity containing man's faith-determined view of reality, aspires to the status of a scientific theory. This status can be granted only if the hypothesis has not been falsified either by observation or by experiment, particularly in the special sciences. Therefore, in view of faith-presuppositions and a hypothesis that is not necessarily the only one possible, it is clear that scientific theories are always *conditioned* and *partial*. They are also relative, for they relate to the knowledge of a particular aspect of reality, such as the physical aspect.

Scientific knowledge, therefore, is achieved by means of a method of analysis and abstraction based on a certain hypothesis. This means, however, that scientific knowledge is also *abstract* and *restricted*. That, of course, is not all, for as conditioned, relative, abstract, and restricted as scientific knowledge may be, it can also grow and even change by varying hypotheses, by a refinement of methods, and by increasing specialization. There is no end to this process of acquiring scientific knowledge.

Concerning the relationship between pre-scientific and scientific knowledge, we can say that the latter must continually be re-embedded in full, direct, concrete human knowledge and experience that has fundamental trust-knowledge as its core. Abstract, scientific knowledge must continually be integrated into and corrected by pre-scientific knowledge, so that the restrictions inherent in scientific knowledge are lifted. In this way, man's integral, pre-theoretical knowledge is gradually enriched and his measure of responsibility is increased.

Reality consists of many more aspects than the one abstracted by a particular science. This must always be kept in mind when the knowledge of one aspect is integrated into pre-scientific knowledge. From a normative standpoint, pre-scientific knowledge requires many-sided enrichment by the integration of numerous forms of scientific knowledge, but this is virtually impossible because of tremendous scientific specialization. It is also important to be open to a multi-disciplinary approach, provided that the sum of various forms of abstract knowledge is never equated with complete knowledge. To achieve this, we need a community of people who can work with one another on the basis of a common view of science. But in this day, with its diffuse religious convictions, this may be extremely difficult to realize.[6]

I shall venture a conclusion at this point. Our fundamental trust-knowledge is the heart of our full and concrete human experience with its factual and applicable knowledge. This pre-scientific experience is enriched by the absorption of new scientific knowledge. It is not only our practical and factual knowledge that profits from this enrichment: our faith-knowledge benefits from it as well. For scientific knowledge is well-suited to the disclosure of more dimensions of the revelation that creation is meant to be. Accordingly, we may say that science brings out truth, but on condition that in all scientific endeavours the Truth is both the starting point and the constant point of return. Thus considered, scientific activity can be carried out with faith as its starting point and a strengthened faith as its result. Then our scientific endeavours will serve the cause of wisdom and of an increasingly comprehensive insight; it will buttress and enlarge our responsibility.

Philosophy of technology[7]

The most important question that ought to occupy us at this time concerns the consequences of this reformational view of science for the development of technology and for a possible redirection of its present development. It will be clear by now that I would be the last person to renounce the scientific basis of modern technology. But I do oppose the *faith in autonomy*, which has combined with science and technology, particularly in western culture. Under its influence, the conduct of science has become the high road to the entire field of knowledge and action, whereas science should be neither more nor less than a helpful supply route.

To gain a better understanding of the auxiliary function of science relative to technology, it is important to note both the motive that ought to drive technological development and its meaning. From God's Word revelation I take the proper motive to be the cultivation and preservation of creation. To limit it to preservation alone would imply a choice for nature and against culture, a renunciation of technology and a submission to natural fate. To limit the motive of technological development to cultivation alone would imply a presumptuousness that fails to consider what is and what is not essential or prudent. God's Word reveals our culture as one that bears the marks of its own extinction within it, threatening both man and nature; hence this Word highlights the cultural destitution we experience.

Within the harmonious calling to cultivate and to preserve, man, the image-bearer of God, is to serve in love. In cultivating and preserving creation he confirms his love towards his Creator and Redeemer and at the same time lovingly represents all of creation. That means, among other things, that man is responsible for unfolding the meaning of creation and that he must resist every attempt to disturb, disintegrate, and destroy this meaning.

Whenever man submits himself to the guidance of this meaningful motive, he will be in a position — *coram Deo* — to accept his task in technology willingly and responsibly. He will pay special attention to the meaning of technology and will attempt to deepen it. For the meaning of technology is rich and manifold; in fact, it is inexhaustible.

Although no one can supply the full meaning of technology, we can state it in part. Technology will be able to alleviate the fate forced on man "by nature." It will offer greater opportunities for living: reducing the physical burdens and strains inherent in labour, diminishing the drudgery of routine duties, averting natural catastrophies, conquering diseases, supplying homes and food, augmenting social security, expanding possibilities for communications, increasing information and responsibility, advancing material welfare in harmony with spiritual well-being, and helping unfold the abundance of individual qualities in people. Moreover, in science and in its own field, technology will develop new possibilities for promoting a variegated disclosure of culture. Technology will also make possible labour that is meaningful as well as productive; it will provide room for work that is marked by creativity, service, love, and care. It will also provide room for leisure and reflection.

This picture of technology, however, is not how it actually

functions today. Inspired by wrong motives, modern technology has been made into a threat to nature, culture, and man; whereas the right motive would lead technology to contribute to the unfolding of nature and to the enrichment and deepening of culture and human life.

To seek both cultivation and preservation, with due respect paid to the meaning of technology, is to pour new and profound content into the extremely high moral purpose of scientists and engineers. For it means that they should no longer arbitrarily follow their own will. Instead, they should eagerly seek to be of service in the unfolding, deepening, and enrichment of meaning. They should not strive to do all things possible, but they should be able to do all things necessary. The purposes, values, and norms for technology should be made explicit in an ethics of technology developed on this basis.

I realize that these observations, too, represent a position contrary to prevailing practice. Very frequently engineers permit themselves to be lured into weighing the advantages. That is what they call ethics. Today such an ethic is impossible, for the scales continue to tip in favour of the disadvantages. It cannot serve to map out the direction technological development ought to take. The actual course and process of development has been accepted as the norm in terms of which a technological ethic weighs the pros and cons. Similarly, no new direction can be expected from an ethic of technology developed after the characteristics of science have first been projected into technology, an act that was inspired by man's determination to gain control and by his concept of technology as applied science.

Ethics based on continued abstraction reduces and restricts ethics itself. It testifies to an urge to continue in a direction that inevitably leads to a dislocation of the meaningful coherence of reality. Although we expect to restore this coherence later without at any point having chosen a new direction, our hope will inevitably remain unfulfilled because of the point of departure. Thus all we can do is fight the symptoms. The cause of the problem will not be eradicated as long as its root is left intact. It is not science and technology that should set up and determine the ethics, for the latter will not be left untouched by the current problems surrounding science and technology. Rather, ethics should *precede* science and technology, so that it may decisively influence their development. The direction technological development will take ought to be determined by our response to ethical issues rather than to technological ones.

A philosophy of technology also requires that an analysis of the relationship between science and technology be given a central position. Such an analysis should examine the basic sciences technology may draw on, the meaning of science for technology, and the precise nature of the technological-scientific method. Ultimately, this philosophical analysis of modern technology must locate the decisive points of contact present in the making of technical designs. That is to say, the engineer's creaturely originality and inventiveness, enriched by a full scientific knowledge, should be assessed as to their place in the designing process.

This done, we need to point out meaningful directions that do justice to the proper motive and meaning of technology. We must also warn against the dangers of wrong motives and wrong choices. As we foster insight into the meaning of technology, we should also seek insight into the dangers and possible nonsense of technology. In this way, the engineer's responsibility for the development of technology is appropriately increased.

Moreover, when the prospective engineer realizes who he is, namely, a human being marked by shortsightedness, shortcomings, and a tendency to underestimate the unfavourable side-effects of his work, he will not be tempted to dominate technological development presumptuously, nor aspire to unlimited achievements. Instead, he will practice wisdom, level-headedness, carefulness, prudence, patience, modesty, and scrupulousness. He will also be prepared to expose his work continually to critique and scepticism. He should desire to interact with his peers in order to find and accept communal responsibility. From that vantage point alone will it be possible to give due attention to the development of technology within the marginal conditions of a historically developed cultural situation and to the necessity for continuity of both the environment and the supply of energy and raw materials.

By emphasizing the responsibility of the engineers, we will be able to slow down current technological development with its ironclad logic and gigantic dimensions and dangers. We will be able to penetrate this development and rework it into a multifaceted, richly varied, and enduring development. There is a tremendous gap between the small-scale technology proposed by the counterculture and the large-scale technology of present culture. It is a gap that needs to be closed in a surprising way by the engineer who seeks to develop the true meaning of technology.

Interaction between technology and culture

In the long run, philosophical reflection on technology should result in an analysis and normative assessment of the societal consequences of technological development. It should also lead us to examine the influence of technology on other cultural sectors. By creating a technostructure, technology forms the basis for other sectors and should contribute to their disclosure instead of to their constriction. Constriction takes place when both the technological-scientific method and the technological results are absolutized instead of serving to disclose various cultural sectors. The meaning of the technostructure, however, is that it serves as a basis for the unfolding and realization of individual and communal responsibilities in areas such as family nurture, education, housing, health services, labour conditions, politics, economics, aid to developing countries, and so forth.

On the other hand, we must also investigate how technology is influenced by culture, particularly from the socioeconomic sector. We should explore the purposes, values, and norms of culture so that we can critically compare them to the normative principles that enhance the meaning of a full cultural life.

The mutual interaction and conditioning of technology and culture constitute a major part of a philosophy of culture. With it we can enrich the engineer's scientific and technological sense of responsibility with a social dimension. We must recognize that other, non-technological points of view ought also to be included in the development of technology.

Finally, keeping in mind the "jolts of our age," in which technology plays a dominant role, I would like to plead for an expansion of the curriculum to enable the student to reflect philosophically on the technological-scientific culture. A thorough study of the origin, motives, and meaning of technology is required, especially for the training of engineers. Without a responsible philosophy of technology, the engineer is likely to remain unaware of his multifaceted responsibility as a bearer of culture. This is even more important now that the technological-scientific culture is faced by nearly insurmountable problems.

In short, a philosophy of technology is essential. It will critically examine the development and problems of technology and the relationship between technology and culture. A philosophical ethics, elaborated on the basis of the proper ethos, must clarify and disclose the motives, goals, values, and norms that guide technological science. In this fashion, we will be able to unfold the originality, creativity, communal spirit, and, above all, the responsibility of the prospective engineer and of all those who are involved in the development of technology.

Notes
1. Cf. Egbert Schuurman, **Technology and the Future** (forthcoming), sections 1.4.2, 2.4.4, 2.4.5, and 2.4.9. Cf. also Chapter 1 section 2 of this book.

2. Cf. H. van Riessen, **Wijsbegeerte** (Kampen: Kok, 1970), sections 4.7.2. (English translation forthcoming).

3. Cf. Schuurman, **op. cit.** , Ch. 2,3 Discussed there are pragmatists and positivists, Norbert Wiener and Karl Steinbuch, the Marxist, Georg Klaus, the transcendentalists, Friedrich Georg Jünger, Martin Heidegger,

Jacques Ellul, Hermann J. Meyer, and the neo-Marxists, Habermas and Marcuse. Cf. also Chapter 1, section 3 of this book.

4. Schuurman, **op. cit.**, section 4.8.

5. Cf. Genesis 11 about God's judgement on the tower of Babel. Cf. also Matthew 4:8-11, a description of the resistance offered by Jesus Christ to the temptation by Satan to take possession of the whole earth, a temptation that is met with radical rejection.

6. Schuurman, **op. cit.**, section 4.6.6.

7. **Ibid.**, Ch. 4.

Reflections on the technological-scientific culture

Select Bibliography

Adorno, Th.W. **Negative Dialektik**. Frankfurt, 1966.

Becher, P. **Mensch und Technik im Denken Friedrich Dessauers, Martin Heideggers und Romana Guardinis.** Frankfurt, 1974.

Beck, H. **Philosophie der Technik — Perspektiven zu Technik — Menschheit — Zukunft.** Trier, 1969.

Bloch, E. **Das Prinzip Hoffnung.** Frankfurt, 1959.

Brinkmann, Donald. "Aufstieg oder Niedergang unserer Kultur?" **Universitas** 2(1947), 1291-1296, 1435-1440.

_____. "Geistige Grundlagen der modernen Technik." **Universitas** 8 (1953), 289-294.

Dooyeweerd, Herman. **Vernieuwing en Bezinning**. Zutphen, 1959. English translation **Reconstruction and Reformation.** Unpublished (available from the Association for the Advancement of Christian Scholarship, 229 College Street, Toronto, Ontario).

Ellul, Jacques. **The Technological Society**. London, 1965.

Gehlen, Arnold. **Die Seele im technischen Zeitalter.** Hamburg, 1957.

Guinness, Os. **The Dust of Death.** Inter-Varsity Press, 1973.

Habermas, Jürgen. **Technik und Wissenschaft als "Ideologie."** Frankfurt, 1968.

Hetman, F. **Society and the assessment of technology.** OECD, 1973.

Heiss, R. **Utopie und Revolution.** Munich, 1973.

Hommes, J. **Der technische Eros — Das Wesen der materialistischen Geschichtsauffassung.** Freiburg, 1955.

Hooykaas, R. **Religion and the Rise of Modern Science.** Edinburgh and London, 1972.

Howe, G. **Gott und die Technik.** Hamburg, 1971.

Hunig, A. **Das Schaffen des Ingenieurs — Beiträge zu einer Philosophie der Technik.** Köln, 1974.

Jungk, R. and J. Galtung, eds. **Mankind 2000.** Oslo and London, 1969.

Kahn, Herman and A.J. Wiener. **The Year 2000**. New York, 1967.

Klaus, Georg. **Kybernetik: Eine neue Universalphilosophie der Gesellschaft?** Berlin, 1973.

Lenk, H., ed. **Technokratie als Ideologie**. Stuttgart, 1973.

_____. **Philosophie im technologischen Zeitalter.** Stuttgart, 1971.

Mekkes, J.P.A. **Teken en motief der creatuur.** Amsterdam, 1965.

Meyer, H.J. **Die Technisierung der Welt — Herkunft, Wesen und Gefahren.** Tübingen, 1961.

Müller-Schwefe, H.R. **Technik und Glaube.** Göttingen, 1971.

Mumford, Lewis. **The Myth of the Machine.** New York, 1967.

_____. **The Transformations of Man.** New York, 1956.

Musgrove, F. **Ecstasy and holiness — Counter Culture and the Open Society.** London, 1974.

Popma, K.J. **Evangelie en geschiedenis.** Amsterdam, 1972. English translation forthcoming.

Rapp, F., ed. **Contributions to a philosophy of technology.** Dordrecht, 1974.

Reich, C. **The Greening of America.** New York, 1973.

Riessen, Hendrik van. **De Maatschappij der Toekomst.** Franeker, 1952; 5th ed. 1974. English translation: **The Society of the Future.** Philadelphia: The Presbyterian and Reformed Publishing Company, 1957.

——————————. **Wijsbegeerte.** Kampen, 1970. English translation forthcoming.

Rohrmoser, G. **Das Elend der Kritischen Theorie.** Freiburg, 1970.

Roszak, Theodore. **The Making of a Counter Culture.** New York, 1969.

——————————. Where the Wastelands Ends. New York, 1972.

——————————. ''The Monster and the Titan: Science, Knowledge and Gnosis.'' **Daedalus.** 1974, 17-32.

Sachsse, H. **Technik und Verantwortung — Probleme der Ethik im technischen Zeitalter.** Freiburg, 1972.

Schelsky, H. **Der Mensch in der wissenschaftlichen Zivilisation.** Koln, 1961.

Schilder, Klaas. **De Openbaring van Johannes en het sociale leven.** Amsterdam, 1924; 4th impr. 1972.

Schumacher, E.F. **Small is Beautiful.** London, 1973.

Schuurman, Egbert. **Techniek en Toekomst.** Assen, 1972. English translation forthcoming.

Senghaas, D. and C..Koch, eds. **Texte zur Technokratiediskussion.** Frankfort, 1970.

Toffler, Alvan. **Future Shock.** New York, 1970.

Waskow, A.I. ''Creating the Future in the Present.'' **The Futurist** 2 (1968), 75 ff.

Weizsäcker, C.F. von. **Tragweite der Wissenschaft.** Stuttgart, 1964.

Whyte, L. ''The historical Roots of our Ecologic Crisis,'' in **Philosophy and Technology.** C. Mitchen and R. Mackey, eds. New York, 1972.

Zuidema, S.U. **De revolutionaire maatschappijkritiek van Herbert Marcuse.** Amsterdam, 1970.